SpringerBriefs in Applied Sciences and Technology

More information about this series at http://www.springer.com/series/8884

Sophie Lufkin · Emmanuel Rey
Suren Erkman

Strategies for Symbiotic Urban Neighbourhoods

Towards Local Energy Self-Sufficiency

Sophie Lufkin
Laboratory of Architecture and Sustainable Technologies (LAST)
Ecole Polytechnique Fédérale de Lausanne
Lausanne
Switzerland

Emmanuel Rey
Laboratory of Architecture and Sustainable Technologies (LAST)
Ecole Polytechnique Fédérale de Lausanne
Lausanne
Switzerland

Suren Erkman
Industrial Ecology Group
University of Lausanne
Lausanne
Switzerland

ISSN 2191-530X ISSN 2191-5318 (electronic)
SpringerBriefs in Applied Sciences and Technology
ISBN 978-3-319-25608-5 ISBN 978-3-319-25610-8 (eBook)
DOI 10.1007/978-3-319-25610-8

Library of Congress Control Number: 2015955383

Springer Cham Heidelberg New York Dordrecht London

Printed on acid-free paper

Springer International Publishing AG Switzerland is part of Springer Science+Business Media (www.springer.com)

Acknowledgments

This book is the result of a research project supported by the programme Collaborative Research on Science and Society (CROSS), jointly set up by Ecole Polytechnique Fédérale de Lausanne (EPFL) and University of Lausanne (UNIL). The authors would like to thank all members of the CROSS programme, as well as all persons who participated in the research project, in particular Félix Burnand, Valérie de Felice and Frédéric Juillard.

Our thanks also go to the representatives of the city of Yverdon-les-Bains, who demonstrated their interest and availability for our research. Stéphane Thuillard, responsible for the Energy Projects, graciously shared his experience with us.

In addition, this publication would not have been possible without the support of many persons whom we would like to thank warmly, specifically all collaborators and partners contributing to research and teaching activities of the Laboratory of Architecture and Sustainable Technologies (LAST) of EPFL and the Industrial Ecology Group of UNIL. The quality of exchanges, the diversity of interdisciplinary inputs and the intensity of the discussions held in these frameworks are precious motivation factors for further developing our research.

More broadly, the authors would like to express their gratitude to EPFL and UNIL, to the Faculties of Architecture, Civil and Environmental Engineering (ENAC-EPFL) and Geosciences and Environment (UNIL) and to the Institute of Architecture (IA-EPFL). Thanks to the quality of their members and their infrastructures, they have provided an appropriate framework for the development of such a publication.

Finally, we would also like to thank Rosemary Besson for her complementary translation, as well as our colleagues, friends and family, who contributed to the successful outcome of the project, thanks to their direct or indirect support.

Contents

About the Authors

Sophie Lufkin architect, graduated from the Ecole Polytechnique Fédérale de Lausanne (EPFL) in 2005. She is the author of a doctoral thesis on the potential for densification of disused railway areas (EPFL, 2010). After an international practical experience at LAR-Fernando Romero (Mexico City), she is currently working as a researcher and lecturer at the Laboratory of Architecture and Sustainable Technologies (LAST) of EPFL, led by Professor Emmanuel Rey.

Her research is part of the global effort towards increased sustainability in the built environment, in particular at neighbourhood scale. She focuses today on the development of an operational indicator system for the integration of sustainability into the design process of urban wasteland regeneration (SIPRIUS project), as well as on design strategies for a new generation of sustainable neighbourhoods called "Symbiotic districts", which are inspired by industrial ecology principles to foster synergies and resource valorization at neighbourhood scale.

Emmanuel Rey earned his architecture degree at the EPFL in 1997, followed in 1999 by a European postgraduate diploma in architecture and sustainable development conferred jointly by the EPFL, the Université Catholique de Louvain (UCL), the ENSA Toulouse and the AA School of Architecture in London. In 2006, he completed his Ph.D. at UCL and was awarded in 2009 the Gustave Magnel prize for the quality of his thesis. Since 2000, he works by the architectural and urban design firm Bauart based in Bern, Neuchâtel and Zurich, of which he became a partner in 2004. In that position, he is involved in many projects, competitions and realizations, which have been published, exhibited and awarded on several occasions. In 2010, he is appointed a professor at the School of Architecture, Civil and Environmental Engineering and founds the Laboratory of Architecture and Sustainable Technologies (LAST).

Emmanuel Rey's contributions focus on the field of sustainable architecture, with special attention to how sustainability principles translate at various levels of the process—from urban design to construction components—and to the incorporation of evaluative and innovative criteria into the architectural project. His interdisciplinary approaches help build dynamic ties between engineers and architects.

Suren Erkman studied philosophy and biology at the University of Geneva and earned his Ph.D. in environmental sciences from the University of Technology at Troyes. After working for several years as a science and business journalist, Suren founded several companies: Institute for Communication and Analysis of Science and Technology (ICAST, Geneva), SOFIES International SA (Geneva, Zurich, Paris and Bangalore), Council on Industrial Ecology (EIC, Paris) and Resource Optimization Initiative (ROI, Bangalore). Suren Erkman is also a professor and head of the Industrial Ecology Group at Faculty of Geosciences and Environment of the University of Lausanne (UNIL).

Abbreviations

AE	Energy reference area
Approx.	Approximately
AW	Area of window
BCE	Before Christian era
BIPV	Building integrated photovoltaics
CE	Christian era
CFF	Chemins de fer fédéraux (Swiss Federal Railways)
CH	Switzerland
CHP	Combined heat and power
CROSS	Collaborative Research on Science and Society
DHW	Domestic hot water
Ed	Editor(s)
EIC	Council on Industrial Ecology
ENAC	School of Architecture, Civil and Environmental Engineering (EPFL)
EPFL	Ecole Polytechnique Fédérale de Lausanne
GFA	Gross floor area
GWh	Gigawatt hour
GWP	Global warming potential
HP	Heat pump
ICAST	Institute for Communication and Analysis of Science and Technology
IMT	Individual motorized transport
ISIE	International Society for Industrial Ecology
kWh	Kilowatt hour
LAST	Laboratory of Architecture and Sustainable Technologies (EPFL)
LMP	Localized master plan
MJ	Megajoule(s)
MW	Megawatt(s)
Non-ren.	Non-renewable
NRPE	Non-renewable primary energy
PEF	Primary energy factor

Pers	Person(s)
PHC	Power–heat coupling
PSC	Parallel studies competition
PV	Photovoltaic
REAP	Rotterdam Energy Approach and Planning
ROI	Resource Optimization Initiative
SEY	Energy Department of Yverdon-les-Bains
SFSO	Swiss Federal Statistical Office
Tot.	Total
TPE	Total primary energy
UCL	Université Catholique de Louvain
UNIL	University of Lausanne
W	Watt(s)
WPP	Water purification plant

Abstract

Within a context of growing efforts to create sustainability strategies, most research in the built environment focuses on energy-related issues. And for good reason, more than one-third of worldwide final energy consumption is attributable to the construction sector. In Switzerland, a landscape dense with urban development, total energy expenditures associated with buildings account for no less than half of the total energy consumption. Ambitious objectives to reduce non-renewable energy consumption are now being set by several European countries, like those developed in Switzerland such as the vision of a 2000 W society or the political decision of phasing out nuclear energy over the medium term.

It was in this spirit that the SYMBIOTIC NEIGHBOURHOODS research project was conceived, in order to simultaneously examine the scientific, technical, urban development and architectural aspects of local energy and resource self-reliance at neighbourhood scale, by integrating issues related to buildings, infrastructure, mobility, food, goods and services. By transposing industrial ecology tools to the field of urban planning, the main objective is to develop new design principles and energy strategies for creating symbiotic neighbourhoods in the Swiss urban context.

Taking lifestyles as a starting point, the research explores three different scenarios (technological, behavioural and symbiotic) for the future development of a neighbourhood for 2035. An energy flow analysis is then achieved, in order to establish an estimated global balance, allowing the assessment and comparison of the energy performance of each scenario. Three quantitative indicators are calculated: total primary energy, non-renewable primary energy and global warming potential. Energy supply, which is defined on the basis of local resources, varies from one scenario to the other. Appreciations on social (acceptability), economic (costs) and environmental (global environmental impact) aspects are also given. They provide a more complete evaluation of each scenario. In parallel, an urban form adapted to the proposed lifestyle is designed, in order to evaluate how architectural and urban design is likely to foster the necessary behaviour changes towards a society consistent with objectives to reduce energy consumption.

The present research report illustrates this approach by analysing the results of a case study on the Gare-Lac sector in the city of Yverdon-les-Bains. The site is currently a large urban Brownfield, intended to host about 3800 additional inhabitants and 1200 jobs upon completion.

The SYMBIOTIC NEIGHBOURHOODS project is jointly led by the Laboratory of Architecture and Sustainable Technologies (LAST) at Ecole Polytechnique Fédérale de Lausanne (EPFL) and the Industrial Ecology Group (Faculty of Geosciences and the Environment, University of Lausanne). It has received funding from the Collaborative Research Programme on Science and Society (CROSS), which aims at encouraging transdisciplinary collaborations between these two institutions.

Keywords Sustainable neighbourhood · 2000 W society · Energy consumption · Industrial ecology · Urban agriculture

Chapter 1
Research Objectives

Abstract The chapter presents the general framework of the “Symbiotic Neighbourhoods” research project. It defines the concept of a symbiotic urban neighbourhood: in addition to increasing the intrinsic efficiency and systematically enhancing renewable resources, the main strategy consists in promoting short cycles and synergies in energy services and material flows. The approach thus allows valorising hidden resources. Within the mature ecosystem resulting from this process, waste produced by one becomes the raw material of the other. Based on this concept, the “Symbiotic Neighbourhoods” research simultaneously examines the scientific, technical, urban development and architectural aspects of local energy and resource self-reliance at neighbourhood scale, by integrating issues related to buildings, infrastructure, mobility and food. By transposing industrial ecology tools to the field of urban planning, the main objective is to develop new design principles and energy strategies for creating symbiotic neighbourhoods in the Swiss urban context.

Keywords Sustainable neighbourhood · 2000-Watt society · Energy consumption · Industrial ecology · Urban agriculture

1.1 Context

Within a context of growing efforts to create sustainability strategies, most research in the built environment focuses on energy-related issues. And for good reason: more than one third of worldwide final energy consumption is attributable to the construction sector (Wallbaum 2012). In Switzerland, a landscape dense with urban development, total energy expenditures associated with buildings account for no less than half of the total energy consumption (Zimmermann et al. 2005). Ambitious objectives to reduce non-renewable energy consumption are formulated today by several European countries, including Switzerland. The concept of the 2000-Watt society, which aims at reducing by a third by 2100 the energy consumption of an average citizen, has become the guiding principle of the national energy policy.

S. Lufkin et al., *Strategies for Symbiotic Urban Neighbourhoods*,
SpringerBriefs in Applied Sciences and Technology,
DOI 10.1007/978-3-319-25610-8_1

In addition, the political decision of phasing out nuclear energy over the medium-term has become the official energy strategy (Eberhard 2004; Previdoli 2012).

At the same time, the projected end of cheap and accessible fossil fuels, geopolitical tensions around the issue of natural resources and the vulnerability of electrical power grids are all factors that motivate the search for strategies allowing access to more secure energy sources, particularly by making use of local resources. With this objective in mind, working toward local energy independence—and specifically a balance between the energy consumption of a territory and its ability to meet its own demand through sustainable production—will tend to minimize environmental impacts while at the same time generating endogenous economic activity and promoting a social and cultural dynamic in which the users can involve themselves. This type of approach requires a significant control of demand (moderation), the widespread use of renewable energy (local production), and an effort to achieve complementarities between operation (industrial symbiosis) (Erkman 1998) and on-site energy storage (Grospart 2009).

Taking these matters into account requires major changes in the way we consider energy in the construction sector by, first of all, clearly transcending the scale of the single building to examine this type of issue at neighbourhood level (Rey et al. 2013). This intermediate scale between the city level and that of the individual building reveals some surprising information, because it takes into consideration the urban reality at a scale that is broad enough to address themes that transcend the single building, while being limited enough to design, test and examine concrete initiatives (Rey 2011). Neighbourhood scale is more restricted—and therefore requires more innovation—than the territories usually considered in most of the current research on local energy independence. It allows taking stock of multi-functionality, considering certain industrial activities and urban or suburban agriculture activities near residential areas, while keeping them to the appropriate scale for the most strategic approaches to urban development (e.g. master plan). This suggests that over the medium term, the results from such a model can be transferred to strategic operational processes.

In parallel, the challenge is also to include a greater number of design parameters, moving well beyond the simple scope of building energy consumption, as the issue is usually addressed in research and practice (heating, domestic hot water, electricity, grey energy). Indeed, recent observation of the functioning of cities suggests that the management of their urban energy flows is not optimal. Their linear metabolism requires substantial quantities of external inputs, to a large extent from non-renewable sources, and generates an important amount of waste, which isn't reused or recycled, such as excess heat or pollution emissions. As Fig. 1.1 shows, this type of linear ecosystem generates limited interactions between the functions.

Because of this suboptimal operation mode, the city attracts, polarizes and generates significant energy flows. The latter increase its environmental impact and could potentially pose the question of its long-term survival.

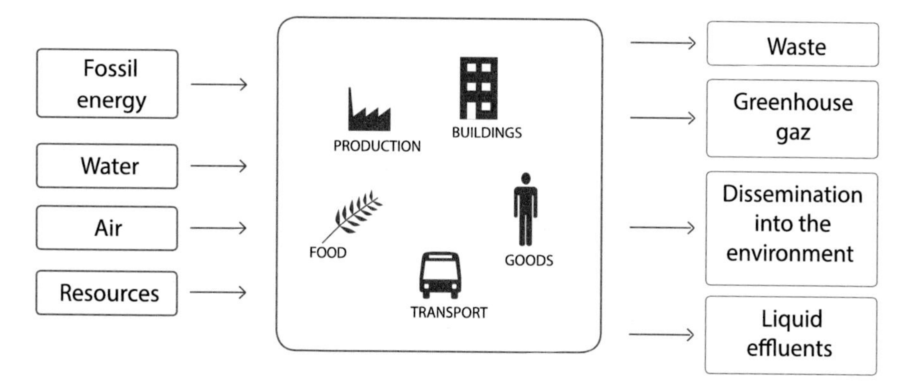

Fig. 1.1 Traditional neighbourhood: linear ecosystem with limited interactions between the functions

1.2 The Concept of a Symbiotic Neighbourhood

In order to work towards greater sustainability, new modalities are therefore required for urban environments to operate in a less dependent and more efficient manner. To this aim, the concept of the symbiotic city seeks to promote a "syntropic" urban system, i.e. an ecosystem that allows the encouragement of the economic and sociocultural development of cities and urban areas, but whose almost cyclical metabolism makes the best use of imported resources and minimises waste rejection. In concrete terms, such an optimisation approach embraces the principles of industrial ecology (Erkman 1998; Faist Emmenegger et al. 2003; Rochat et al. 2006; Erkman and Massard 2011). It is based on three complementary and convergent axes: (1) increasing the intrinsic efficiency of cities, (2) systematically enhancing renewable resources and (3) implementing urban symbioses.

First of all, increasing the intrinsic efficiency of a city involves reducing its needs, in particular by means of a greater coordination between urbanisation and transport policies. The concept of a polycentric compact city, which implies promoting urban densification in close proximity to public transport and making the best use of untapped potentials in the built environment, is fully in line with this objective. However, due to the complexity of interactions in the urban environment, it should be noted that an action on the sole densification would be simplistic and insufficient (Rey 2011). Indeed, densification is not the unique remedy to urban planning issues. Therefore, reducing the city's needs goes way beyond considerations of buildings localisation or urban compactness. It integrates multiple spatial, environmental, sociocultural and economic objectives, such as increasing energy efficiency in buildings (compactness of the urban form, high-performance thermal insulation, intelligent building control), minimising losses in networks (drinking water, heating, electricity) or promoting responsible consumption behaviours (energy, drinking water, food, goods and services).

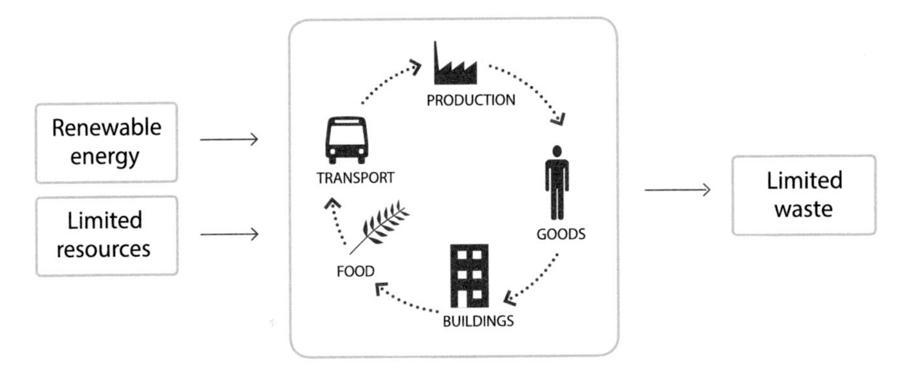

Fig. 1.2 Symbiotic neighbourhood: mature ecosystem with intensive resources cascading and reuse

The second optimisation axis is the systematic integration of renewable resources into the functioning of cities, based on targeted and adequate use of available local resources. Among the various fields considered, the energy sector occupies a central positioning. Many strategies can be implemented: hydraulic, geothermal and solar energy (passive systems, solar thermal and solar electric panels) and biomass.

The third optimisation axis is the encouragement of urban symbioses (Lufkin et al. 2013a). Promoting short cycles and synergies in energy services and material flows, the approach consists in particular of valorising hidden resources at neighbourhood, city or urban region scale. Within the mature ecosystem resulting from this process, waste produced by one becomes the raw material of the other (see Fig. 1.2).

1.3 State of the Art

The idea of considering the city as an ecosystem was introduced in the 60s. Biology and the theory of the ecosystems are used as sources of inspiration to deal with the complex reality and to perceive the environment in a more systemic approach. Urban metabolism thus provides a way to analyse resource and material flows associated with cities (Baccini 1996). With this rigorous basis, researchers proceeded to the study of several cities (Newman 1999), highlighting the malfunctions of the urban metabolism: high dependence on fossil energy sources, low efficiency due to the linear processes, ineffective sectoral policies, end of pipe solutions, etc. (Barles 2008; Dobbelsteen et al. 2012). Urban metabolism represents a very efficient tool to assess the sustainability of a city and to identify potential resources and waste that could be reused at regional level (Codoban and Kennedy 2008).

Yet, the scale of the city remains too large to be operational in terms of urban design. The transition between analytical and prospective approaches thus remains problematic. Only a limited number of experiences aiming at integrating the concept of urban symbioses at neighbourhood scale have been carried out. These pilot projects have been inspirational for this research project. They include the *REAP Methodology*, in Rotterdam (Tillie et al. 2009), the *Amsterdam Guide to Energetic Urban Planning* (Tillie et al. 2011), the *Urban Harvest Concept* in Kerkade West (Agudelo-Vera et al. 2012) or the *New Stepped Strategy* (Dobbelsteen 2008). All thee pilot projects are still at the experimental or planning stage; at the present time, none has reached completion.

The integration of symbiotic strategies at building scale, just like at neighbourhood scale, is at its infancy. Indeed, the most widespread strategies to deal with the energy challenge focus on reducing the demand. They include measures to increase the intrinsic efficiency of the building (compactness of the form, excellent thermal insulation, intelligent regulation), the minimisation of network losses (drinking water, heating, electricity) or the promotion of moderate behaviour of users (Lufkin et al. 2013b).

Yet, some recent realizations have demonstrated the powerful potential of a more systematic integration of renewable resources into architectural projects. The Microcity complex (Veillon 2012), located in Neuchâtel and recently built for EPFL, highlights how a building can contribute to the energy supply of an urban sector that clearly exceeds the limits of a single building. Its architecture integrates both a photovoltaic solar roof (linked to the urban grid) and a cooling loop functioning with water from the nearby lake (also used by other equipment buildings of the neighbourhood).

On another level, other projects highlight that the implementation of urban symbioses can become a reality only if buildings themselves start to move beyond simple mix-use and to pragmatically combine unexpected functions. In this sense, this approach leads to the emergence of new types of buildings, which we refer to as "neotypes". Culture, leisure, housing, production and agriculture can live together within the same building and benefit from the proximity to one another.

In this perspective, public infrastructure and complex equipment buildings such as transport hubs, sports facilities, laboratories, institutions (the Microcity building, for instance), because of their more perennial and structuring nature, appear as opportunities to foster the transition towards a symbiotic architecture. The project developed by architect Pascal Gontier for the aquatic sports centre within Paris' Olympic bid in 2012 (Rey 2013) provides another relevant illustration of how hybridisation of the functions creates added value both for the building and for its surrounding neighbourhood. The urban symbiosis occurs between the natural Olympic swimming pool, greenhouses and the mixed-use nearby neighbourhood. The lush garden of the greenhouses serves several purposes: solar energy valorisation—passive strategy in the greenhouses with building integrated photovoltaics (BIPV) on the roof—and redistribution of excess heat produced to the surrounding buildings. In parallel, through a process of phytoremediation, the plants purify the pool water; the garden is used as a neighbourhood pleasure park; vegetable gardens provide local and seasonal products to the inhabitants.

These public projects can symbolically play a triggering role to promote the application of symbiotic principles within more private buildings, such as commercial centres, industrial techno-parks, or large hotels which, due to their dimensions, are likely to generate the same short distance virtuous cycles. In Rotterdam, within the frame of an analysis of the synergy potentials between buildings, these hybrid buildings have been termed "multifunctional clusters" (Tillie et al. 2009). One of the projects foresees a building with activities on the lower floors (offices, businesses, ice-rink) and community housing on the upper floors with photovoltaic (PV) rooftop. The diversity of energy demand allows optimizing the use of the captured solar energy, while residential functions benefit from the waste heat produced by the activities.

Although not yet completed, these various examples demonstrate the relevance and realism of symbiotic approaches. It remains for us today to integrate them into coherent reflexion at the neighbourhood level in order to influence the energy supply in an urban sector which would extend far beyond the physical limits of a single building.

As the operational practices in the urban planning and construction sectors are characterized by a certain conservatism, a rigorous and scientific demonstration of the feasibility and realism of integrating symbiotic approaches at neighbourhood scale is needed. It is precisely the ambition of the present research project.

1.4 Objectives of the Research

The symbiotic neighbourhood concept thus resides in the will to transcend the limits of sectorial approaches by combining both the parameters linked with buildings, the infrastructures, mobility, food and the consumption of goods and services. To date, there have been very few studies that include these aspects with such a broad perspective that address energy supply in particular as both an energy consideration—even though power supply accounts for a significant portion of the total energy balance per inhabitant (Rey 2006)—and from an urban development standpoint—urban agriculture is becoming an increasingly popular consideration with regard to achieving urban sustainability (Gorgolewski et al. 2011; Jourdan and Mirenowicz 2011).

It was in this spirit that the "Symbiotic Neighbourhoods" research project was conceived, in order to simultaneously examine the scientific, technical, urban development and architectural aspects of local energy and resource self-reliance at neighbourhood scale, by integrating issues related to buildings, infrastructure, mobility and food. Concretely, the main objective is to transpose the use of industrial ecology tools (Erkman 1998; Faist Emmenegger et al. 2003; Rochat et al. 2006; Erkman and Massard 2011) to the urban environment. Inspired by the concept of urban symbiosis, the project aims at developing design principles and energy strategies for creating symbiotic neighbourhoods in the Swiss urban context.

The research goal is also to identify the most efficient means of leverage to reduce energy consumption—lifestyles, technologies or urban form—and study the interactions between these three lines of intervention. Indeed, in spite of growing awareness on the part of private or public stakeholders regarding energy-related issues, they are slow to put this into practice and adopt responsible behaviour, sometimes because of a lack of available relevant information. By establishing reliable bases for addressing the energy question in sustainable urban neighbourhoods in the future (horizon 2035), the suggested approach thus enables systematic exploration of the links—as yet still to be created—between strictly quantitative issues related to energy efficiency (based on industrial ecology) and the qualitative and operational issues linked with their implementation in approaches being used in actual projects (Erkman 1998).

The results presented in this research report focus more specifically on a case study on the Gare-Lac sector of the town of Yverdon-les-Bains (Figs. 1.3 and 1.4), currently a large urban brownfield, and intended to host some 3800 inhabitants and provide 1200 extra jobs in a new neighbourhood covering approx. 23 ha. The approval procedure of the Localised Master Plan (LMP) of this vast area is in progress (Fig. 3.4). The document is currently at the public consultation phase (Bauart et al. 2013).

Fig. 1.3 Orthophotograph of the Gare-Lac sector of Yverdon-les-Bains highlighting the Gare-Lac area, which corresponds to a large urban brownfield strategically located between the station and the Lake of Neuchâtel

Fig. 1.4 Aerial photograph of the Gare-Lac sector in Yverdon-les-Bains. The site is an archetype of urban brownfield: it can be defined by a combination of derelict buildings, vacant lots and abandoned factories. It is crossed by three canals and marked by proximity to the Lake of Neuchâtel (*Source* LAST, August 2012)

References

Agudelo-Vera CM, Leduc WRWA, Mels AR, Rijnaarts HHM (2012) Harvesting urban resources towards more resilient cities. Resour Conserv Recycl 64:3–12. doi:10.1016/j.resconrec.2012.01.014

Baccini P (1996) Understanding regional metabolism for a sustainable development of urban systems. ESPR Environ Sci Pollut Res 3:108–111

Barles S (2008) Comprendre et maîtriser le métabolisme urbain et l'empreinte environnementale des villes. Responsab Environ N° 52:21–26

Bauart et al (2013) Plan Directeur Localisé Gare-Lac. Commune d'Yverdon-les-Bains, Yverdon-les-Bains

Codoban N, Kennedy CA (2008) Metabolism of neighborhoods. J Urban Plan Dev 134:21–31

Eberhard J (2004) Steps towards a sustainable development a white book for R&D energy-efficient technologies. Novatlantis

Erkman S (1998) Vers une écologie industrielle. Charles Léopold Mayer, Paris

Erkman S, Massard G (2011) Ecologie industrielle à Genève. Le transport de marchandises. Enjeux pour Genève, Geneva

Faist Emmenegger M, Cornaglia L, Rubli S (2003) Métabolisme des activités économiques du canton de Genève—Phase 1. Geneva

Gorgolewski M, Komisar J, Nasr J (2011) Carrot city: creating places for urban agriculture. New York City

Grospart F (2009) Autonomie énergétique locale. Ecocentre Habitat, Vendôme

Jourdan S, Mirenowicz J (2011) L'agriculture regagne du terrain dans et autour des villes. La Rev Durable 43:15

Lufkin S, Rey E, Erkman S (2013a) Symbiotic districts—innovative design strategies for local energy and resource self-reliance at the district scale by integrating issues related to buildings, infrastructures, mobility and food. In: 7th international society for industrial ecology Biennal conference (ISIE 2013)

Lufkin S, Rey E, Erkman S (2013b) Quartiers symbiotiques: augmenter le potentiel d'autonomie énergétique à l'échelle locale. In: Vers la ville symbiotique? Valoriser les ressources cachées. 7ème édition du Forum Ecoparc, Stämpfli. Tracés, Bern, pp 16–19

Newman P (1999) Sustainability and cities : extending the metabolism model. Landsc Urban Plan 44:219–226

Previdoli P (2012) Energiestrategie 2050. Bundesamt für Energie BFE, Bern

Rey E (2006) Integration of energy issues into the design process of sustainable neighborhoods. In: PLEA2006—the 23rd conference on passive and low energy architecture. Geneva

Rey E (2011) Concevoir des quartiers durables. In: OFEN/ARE (ed) Quartiers durables. Défis et potentialités pour le développement urbain, OFEN/ARE. Bern, pp 15–24

Rey E (2013) Vers la ville symbiotique. In: Vers la ville symbiotique? Valoriser les ressources cachées. 7ème édition du Forum Ecoparc. pp 3–5

Rey E, Lufkin S, Renaud P, Perret L (2013) The influence of centrality on the global energy consumption in Swiss neighborhoods. Energy Build 60:75–82

Rochat D, Erkman S, Chambaz S (2006) Le recyclage des matériaux de construction à Genève. République et Canton de Genève, Geneva

Tillie N, Van Den Dobbelsteen A, Doepel D et al (2009) Towards CO_2 neutral urban planning: presenting the rotterdam energy approach and planning (REAP). J Green Build 4:103–112. doi:10.3992/jgb.4.3.103

Tillie N, Kürschner J, Mantel B, Hackvoort L (2011) The Amsterdam guide to energetic urban planning. Amsterdam

van den Dobbelsteen A (2008) Towards closed cycles—new strategy steps inspired by the Cradle to Cradle approach. In: PLEA 2008—25th conference on passive and low energy architecture. Dublin

van den Dobbelsteen A, Keeffee G, Tillie N, Roggema R (2012) Cities as organisms. In: Roggem R (ed) Swarming landscapes: the art of designing for climate adaptation. Springer, Dordrecht, pp 196–206

Veillon E (2012) Microcity : un projet hybride en bois-béton. batimag 1:6–10

Wallbaum H (2012) Mainstreaming energy and resource efficiency in the built environment—just a dream ? In: IED public lecture series (ed). IED public lecture series

Zimmermann M, Althaus H-J, Haas A (2005) Benchmarks for sustainable construction. Energy Build 37:1147–1157. doi:10.1016/j.enbuild.2005.06.017

Chapter 2
Methodology

Abstract The chapter presents the methodology implemented for reaching the objectives of the "Symbiotic Neighbourhoods" research project. Taking lifestyles as a starting point, the research explores three different scenarios (technological, behavioural and symbiotic) for the future development of a neighbourhood for 2035. An energy flow analysis is then achieved, in order to establish an estimated global balance, allowing the assessment and comparison of the energy performance of each scenario. Three quantitative indicators are calculated: total primary energy, non-renewable primary energy and global warming potential. Energy supply, which is defined on the basis of local resources, varies from one scenario to the other. They provide a more complete evaluation of each scenario. In parallel, an urban form adapted to the proposed lifestyle is designed, in order to evaluate how architectural and urban design is likely to foster the necessary behaviour changes towards a society consistent with objectives to reduce energy consumption.

Keywords Sustainable neighbourhood · 2000-Watt society · Energy consumption · Industrial ecology · Urban agriculture · Energy mapping · Flow analyses

2.1 Energy Mapping

The research focuses on four distinct phases.

First of all, in order to estimate available local resources, the first stage consists in establishing a regional energy map. In the first Instance, on the basis of an earlier report (Weinmann Energies 2010) and using information obtained from the Energy Department of Yverdon-les-Bains (SEY), the list of the different renewable energy production installations present in 2013 and located within the perimeter of the Yverdon-les-Bains conurbation was established. Energy supply projects for the municipality of Yverdon-les-Bains planned by the SEY up until 2035 were also assessed.

S. Lufkin et al., *Strategies for Symbiotic Urban Neighbourhoods*,
SpringerBriefs in Applied Sciences and Technology,
DOI 10.1007/978-3-319-25610-8_2

These projects are essentially based on local resources potential. i.e. biomass, the multiple types of waste that can be recycled either using a process of biomethanisation, or to produce heat through combustion, but also the local rate of sunshine for solar power, as well as geothermal energy and wind power. This information enabled researchers to draw up a mapped inventory of the various local resources available.

During the final stage of establishing this energy registry, the different potentially promising zones for heat recovery from waste heat were mapped. The global overview of these results is presented in Chapter 4.1 "Local resources".

2.2 Scenarios

On the basis of this inventory and using the Localised Master Plan (LMP) as the frame of reference, three scenarios for the neighbourhood development with 2035 as the horizon were imagined, fuelled by prospective thinking on the evolution of western lifestyles (Emelianoff et al. 2012). Each scenario deliberately corresponds to a distinct vision of how to reach sustainability objectives and thus, in a way, embodies a different intellectual point of view:

- The *technological scenario* calls on cutting-edge technologies to reduce energy consumption and emission of greenhouse gas. No substantial user-behavioural change is expected and the current trend continues. Globally, environmental impacts are lessening thanks to improved efficiency of control mechanisms, but this effect is counterbalanced by the general increase in consumption.
- The *behavioural scenario*, on the other hand, is quite the opposite of the technological scenario, relying mainly on a meaningful evolution in users' behaviour: lower consumption, voluntary simplicity, lower consumerism and a 'return to favour' of slowness. In this scenario the driving force behind energy transition thus depends essentially on a reduction in the final demand thanks to a change in certain current social practices.
- The *symbiotic scenario* gives even higher value to the possibilities of urban and industrial symbiosis to reduce the neighbourhood's impact on the environment. It aims in particular to transform lost waste (particularly unavoidable heat production) into resources. The approach implies maximising the use of exchange networks for material and energy, at all levels (buildings, groups of buildings, neighbourhood, between the neighbourhood and its surrounding perimeter). With respect to the behavioural dimension, this scenario is mid-way between the previous two, as the users will act according to the logics of networks and partnerships.

Table 2.1 provides a synthetic overview of the hypotheses made for the technological, behavioural and symbiotic scenarios according to five energy consumption sectors.

Table 2.1 Overview of the hypotheses made for the technological, behavioural and symbiotic scenarios according to five energy consumption sectors (buildings, mobility, infrastructures, food, goods and services)

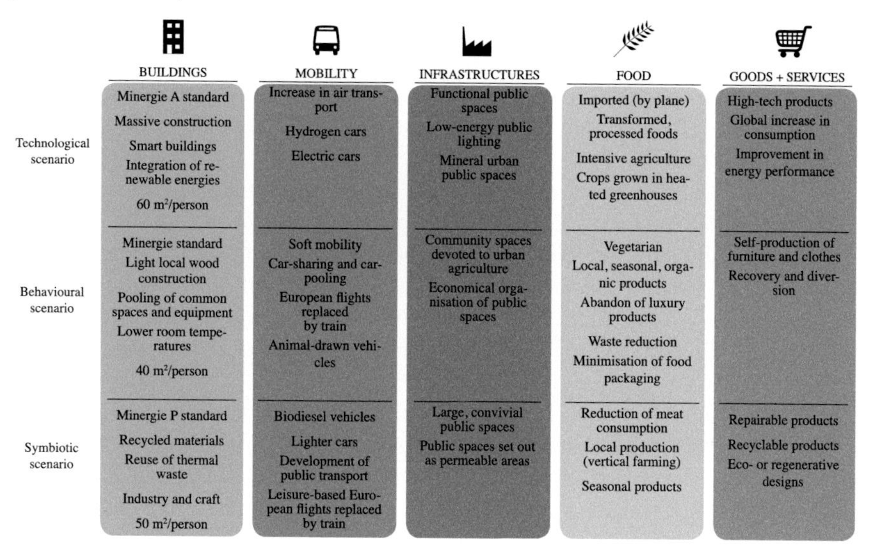

	BUILDINGS	MOBILITY	INFRASTRUCTURES	FOOD	GOODS + SERVICES
Technological scenario	Minergie A standard Massive construction Smart buildings Integration of renewable energies 60 m²/person	Increase in air transport Hydrogen cars Electric cars	Functional public spaces Low-energy public lighting Mineral urban public spaces	Imported (by plane) Transformed, processed foods Intensive agriculture Crops grown in heated greenhouses	High-tech products Global increase in consumption Improvement in energy performance
Behavioural scenario	Minergie standard Light local wood construction Pooling of common spaces and equipment Lower room temperatures 40 m²/person	Soft mobility Car-sharing and car-pooling European flights replaced by train Animal-drawn vehicles	Community spaces devoted to urban agriculture Economical organisation of public spaces	Vegetarian Local, seasonal, organic products Abandon of luxury products Waste reduction Minimisation of food packaging	Self-production of furniture and clothes Recovery and diversion
Symbiotic scenario	Minergie P standard Recycled materials Reuse of thermal waste Industry and craft 50 m²/person	Biodiesel vehicles Lighter cars Development of public transport Leisure-based European flights replaced by train	Large, convivial public spaces Public spaces set out as permeable areas	Reduction of meat consumption Local production (vertical farming) Seasonal products	Repairable products Recyclable products Eco- or regenerative designs

2.3 Flow Analyses

For each scenario, several hypotheses are then formulated bearing in mind the energy issues relating to the fields of buildings, mobility, infrastructures, food and consumption of goods and services. The energy consumption of each of these scenarios is calculated on the basis of these hypotheses, using analysis of energy flow. Three quantitative indicators have been retained: total primary energy, non-renewable primary energy and global warming potential (emissions of CO_2 equivalent).

An initial calculation bears in mind the current situation, representative of the status of the neighbourhood if it were inhabited to full capacity. This current status is a point of reference which has subsequently enabled adaptation of data depending on the hypotheses developed for each scenario. These calculations are made in detail for each area and every category of energy consumption. The hypotheses used to build up the scenarios are summed up in Table 1.1. They are addressed in more depth in the annexes to this report.

The first area analysed is *Buildings*. It includes the residential and office categories which each comprise sub-categories on Construction, Heating, Domestic Hot Water (DHW), Ventilation and Lighting, and Appliances. This section considers the building's grey energy as well as the electricity and heat used during its operational period. Calculation is based on the different standards of Minergie control (Minergie, Minergie A and Minergie P). These reference values, per square meter, are then applied to the residential and office surface areas, per inhabitant, in the scenario considered.

The *Mobility* section includes the following categories: Car, Plane, Train and other Public Transport. This data comes from the listings used by the Swiss Federal Statistical Office (SFSO). Only passenger transport is included here, goods transport is allocated to 'goods'.

Systematically, the *Infrastructures*' final energy consumption is estimated in accordance with a pro rata principle of the populations concerned. The first category comprises Equipment, i.e. both the neighbourhood facilities (whose total consumption is allocated to the inhabitants of the Gare-Lac sector), communal facilities, such as swimming pool, ice-rink, etc. (of which the neighbourhood residents pay only 14 %, i.e. the ratio between the neighbourhood and the town of Yverdon-les-Bains) and finally regional facilities provided for the whole conurbation of Yverdon-les-Bains (St-Roch, CFF workshop), of which only 7 % are borne by residents of the neighbourhood. The Remaining Infrastructures category comprises the grey energy of these facilities (its estimation would deserve a complete analysis although it remains constant from scenario to scenario) as well as the grey energy linked to infrastructures servicing the residents (roads and refuse department, small-scale installations) which do not vary according to the scenarios. The final category External Installations corresponds to the higher pace consumption represented by the large domestic producers/transmitters, i.e. infrastructures releasing over 15,000 tonnes of CO_2 eq/per year (Bébié et al. 2010). This national data is shared across the whole of the Swiss population to obtain an average figure per inhabitant.

The fourth area bears in mind energy and greenhouse gas emissions related to *Food*. It includes agriculture, processing, packaging and distribution. Estimations are based on recent surveys which evaluate the primary energy consumption at approx. 750 W per inhabitant (Jungbluth and Itten 2012). Categories with the most interesting potential for reducing energy consumption are then identified which enables researchers to vary results from one scenario to another.

The last area accounts for *Goods and Services*-related energy used by households, which does not feature in any other area's calculations. Products with a short lifespan (clothes, furniture, etc.) and non-standard activities (concerts, hotel stays, etc.) are the highest grey energy consumers (Novatlantis 2011). Currently, the electrical power linked with goods and services consumption is evaluated at approx. 750 W per person. Just as for the area of *Food*, each scenario adapts this estimation of the current status of *Goods and Services*-related energy consumption according to the hypotheses made.

Once the three indicators are calculated for the different areas, the values are summed up in order to estimate the overall assessment of each scenario. To check the relevance of the reflection, these results are then compared with objectives set out in the framework of a 2000-Watt society for 2035.

2.4 Urban Form

In parallel, an urban form proposal adapted to the lifestyle described in the scenario is modelled. This spatial installation allows researchers to determine the neighbourhood's hosting capacity. Indeed, each lifestyle implies distinct usages corresponding to specific needs in terms of spatial, functional and sensitive qualities (Thomas 2011).

This foresight exercise in translating conceptual hypotheses into a built-up form also enables us to evaluate the extent to which the architectural and urban quality is likely to support—even encourage—the behavioural changes required for the transition towards a society compatible with the reduced energy consumption goals.

The development scenarios use the Localised Master Plan (LMP) prepared for the Gare-Lac sector of Yverdon-les-Bains (Bauart et al. 2013) as their reference framework. The specific identity of the future neighbourhood and its closeness to the historical town centre, the railway station and the lake all contribute to making it a rare site in the urban environment. The aim is to make the most of these advantages not by densifying at all costs, but rather by introducing mixed-use planning and a balance between the built and non-built spaces. Established by the Localised Master Plan (LMP), human density (the number of inhabitants and jobs per hectare) remains identical for all the scenarios. On the other hand, the built density (gross floor area) and the urban form vary from one scenario to another, depending on the function of the area per inhabitant and the type of activities focused on in the different scenarios. Based on different sustainable development visions, the scenarios aim to tangibly explore various forms of urban development and siting of buildings.

2.5 Optimisation

Finally, the last phase involves optimizing the scenarios' performances by developing a scenario incorporating the three visions. The hypotheses retained for this hybrid scenario thus reflect a less radical and more pragmatic and realist vision of sustainable development. At this stage, the flow analysis results are compared once again with the goals of the 2000-Watt society.

Figure 2.1 presents a diagram of research progress. In spite of the apparent linearity of the process, numerous iterations are made between the different stages in research.

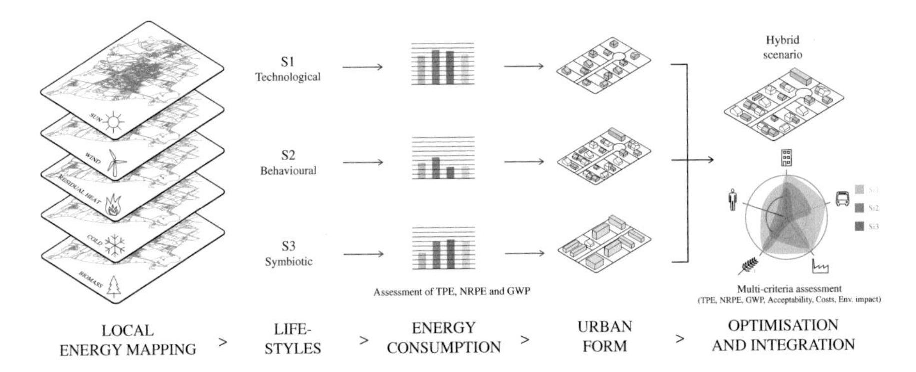

Fig. 2.1 Methodological diagram describing the different stages in the symbiotic neighbourhoods research project process

References

Bauart et al (2013) Plan Directeur Localisé Gare-Lac. Commune d'Yverdon-les-Bains, Yverdon-les-Bains

Bébié B, Hartmann C, Lenel S et al (2010) Les cités de l'énergie, les villes, les communes et les régions sur la voie de la société à 2000 watts. Cossonay

Emelianoff C, Mor E, Dobre M et al (2012) Modes de vie et empreinte carbone. Prospective des modes de vie en France et empreinte carbone, Paris

Jungbluth N, Itten R (2012) Umweltbelastungen des Konsums in der Schweiz und in der Stadt Zürich: Grundlagendaten und Reduktionspotenziale. Zurich

Novatlantis (2011) Vivre plus légèrement. Vers un avenir énergétique durable: l'exemple de la société à 2000 watts. Villigen

Thomas M-P (2011) En quête d'habitat : choix résidentiels et différenciation des modes de vie familiaux en Suisse. EPFL

Weinmann Energies (2010) Parc scientifique et technique d'Yverdon-les-Bains (Y-Parc). Etude pour l'évaluation des rejets thermiques et des besoins en chaleur du site. Echallens

Chapter 3
Case Study

Abstract The chapter presents the case study on which the methodology of the "Symbiotic neighbourhoods" research project was applied. It focuses on the Gare-Lac neighbourhood in Yverdon-les-Bains (Switzerland), a section of town which is particularly emblematic of the challenges linked with urban densification. Currently comprising numerous brownfield plots, the sector to be developed represents an area of 23 ha ideally located between the railway lines and the Lake of Neuchâtel. A new Localised Master Plan (LMP) is soon to become operative; it foresees the arrival of 3800 additional inhabitants and 1200 jobs. On this basis, three different scenarios (technological, behavioural and symbiotic) have been elaborated for the future development of the neighbourhood.

Keywords Sustainable neighbourhood · 2000-Watt society · Energy consumption · Industrial ecology · Urban agriculture · Gare-Lac sector · Localised master plan Gare-Lac

3.1 Presentation of the Site

The case study focuses on the Gare-Lac neighbourhood in Yverdon-les-Bains. This section of the town is particularly emblematic of the challenges linked with urban densification. Currently comprising numerous brownfield plots, the sector to be developed represents an area of 23 ha ideally located between the railway lines and the Lake of Neuchâtel.

Like in so many other cities, the urban development of Yverdon-les-Bains is strongly linked with transportation arteries and the Lake of Neuchâtel. As ideal hubs for trading and exchange, numerous towns indeed grew up on these crossroads with a view to taking advantage of the favourable possibilities for travel and goods transport as well as food and energy production, and the leisure activities provided by seafronts, lakes and rivers. Yverdon-les-Bains, whose name is intrinsically linked with water, is no exception to this rule. Located on the western shore of the

S. Lufkin et al., *Strategies for Symbiotic Urban Neighbourhoods*,
SpringerBriefs in Applied Sciences and Technology,
DOI 10.1007/978-3-319-25610-8_3

Lake of Neuchâtel, at the mouth of the river Thièle, it enjoys a privileged position on the Swiss plateau.

The creation of a new neighbourhood in the Gare-Lac sector is in line with its historic ties. It provides a unique opportunity for the town to reorient its urban development towards the interior and reclaim its urban lakeside areas. Following a short outline of the history of Yverdon-les-Bains, this chapter seeks to present the emblematic site's main redevelopment issues, as a true archetype of redundant urban industrial land, crossed by three canals and influenced by the immediate vicinity of the Lake of Neuchâtel.

3.1.1 An Urban Settlement Influenced by the Vicinity of the Lake

The first settlers built their dwellings along the shores of the Lake of Neuchâtel from about 4000 BCE (Aubert 1995). These were the riverside dwellers who erected a series of stone megaliths in the Baie de Clendy, protected as a UNESCO World Heritage "pile dwelling" site since 2011, which lies just outside the Gare-Lac perimeter at the eastern end of the Avenue des Sports (Fig. 3.1).

At the beginning of the first millennium BCE, the first inhabitants gradually abandoned their lakeside settlements no doubt due to a disastrous rise in lake water levels (Kasser 1969). The arrival of the Helvetians a few centuries later instigated the construction of a fortified town called Eburodunum on both banks of the river Thièle. Under Roman domination, urbanisation of the region flourished. By that time, Eburodunum's population had swelled to about a thousand soldiers, craftsmen and boatmen.

During the second half of the first millennium CE, the shores of the lake receded northwards by several hundred meters. A new downstream shoreline stranded Eburodunum upstream and led to its gradual abandonment. It also brought about the construction of the medieval city and the castle on the new shore (Weidmann 2014). In the Middle Ages, Yverdon-les-Bains was a military garrison town built on a gentle rise of land surrounded by waterways—the various branches of the Thièle—the lake, to the north, and marshland. Spreading from the castle and surrounded by ramparts (Fig. 3.1), the town was rapidly organized around the three main streets which still dominate the old town's historic layout today (Weidmann 2014). The city was the obligatory thoroughfare across dry land and as a major staging-post between lake and river traffic, it built its prosperity on the strategic position which enabled it to control traffic and trade. And yet this same position, isolated by water, side-lined it from the main trade routes in the early 19th century, although the town was actually located on the shortest route between Geneva and Bern or Zurich.

Fig. 3.1 Historical evolution of the city of Yverdon-les-Bains. Since its establishment and until the middle of the 19th century (*top left image*), the city seeks proximity of water. Later on, the major hydrographic and railway infrastructures involve a period of disinterest, which corresponds to the substitution of waterway transport by terrestrial transport and to the negative image associated to flood risk (*top right*). The railway cut off the old medieval city from the lake while construction of the Swiss Federal Railway (CFF) workshops launched the industrialization of reclaimed land (*bottom left*). The new Localised Master Plan (LMP) Gare-Lac reflects a spirit of creating new alliances between the lake and the contemporary city (*bottom right*). *1* Prehistoric site; *2 Castrum*; *3* Medieval town; *4* Castle; *5* Ancient port; *6* Station; *7* CFF workshops; *8* Perimeter of the LMP Gare-Lac; *Dotted lines*: Evolution of the lake shore

3.1.2 Early Industrialization: Physical and Functional Dissociation

Things changed radically with the first Jura water correction (1868–1891), which lowered the lake's average water level by over 2 m and created 130 ha of new land between the town and the lake (Nast 2006). Built on marshland, the railway cut off the old medieval city from the lake while construction of the Swiss Federal Railway (CFF) workshops launched the industrialization of reclaimed land (Fig. 3.1).

The industrial boom, substantiated by the arrival of new enterprise (metallurgy, food industry, construction), was characterized by the expansion of mechanisation and machinery no longer relied on extensive use of water as its driving force. Hydraulic power was replaced by the steam engine, which in turn was superseded by the first gasworks and then electricity power stations (De Raemy and Auderset 1999). This meant that industrial buildings were not necessarily built along waterways, but along lines of communication, particularly railways. Finally, the channelling of the river Thièle and the disappearance of the bustling port activity brought about a permanent, functional and geographical split between the old town and its lake- and riverside neighbourhoods (Fig. 3.1).

Comparison of the 1852 map, before the construction of major hydrographic and railway infrastructures, and the current situation (Fig. 3.1) reveals the profound changes that have affected the area. In those days, numerous canals still evacuated marsh water and the course of rivers differed significantly from their current trajectory.

3.1.3 *Towards Urban Reconciliation?*

In 1902 the municipal authorities organized a competition for the town's First Extension Plan, stressing that shores of the lake were to be developed for leisure purposes. Very few roads were actually built. In the early 80s the municipality abandoned the idea of a main road along the lakeside which had been planned from the 1950s (De Raemy and Auderset 1999).

In the late 20th century, as a result of industrial decline and the tertiarisation of activities, many businesses such as Hermès Precisa International, Leclanché, Arkina and Vaucher ceased production in Yverdon-les-Bains, and activities at the military arsenals and the slaughterhouses came to a standstill.

At the same time, vast wilderness areas flourished on the land that had emerged after the water correction, such as the Grande-Cariçaie, for example, which is the largest reed bed in Switzerland and an internationally recognized natural site.

3.2 Demographic Projections

Today, with its 25,000 inhabitants, Yverdon-les-Bains plays several key roles by being simultaneously the second largest town in the canton of Vaud (ranking 24th in Switzerland), the centre of a flourishing conurbation (with the aggloY project) and the hub of the vast Nord Vaudois region. Its strong features are its high quality of life enhanced by its historical heritage and geographical advantages such as the magnificent Jura landscapes and immediate proximity of the Lake of Neuchatel. Being a hub like this is also beneficial on the cultural, educational, commercial and

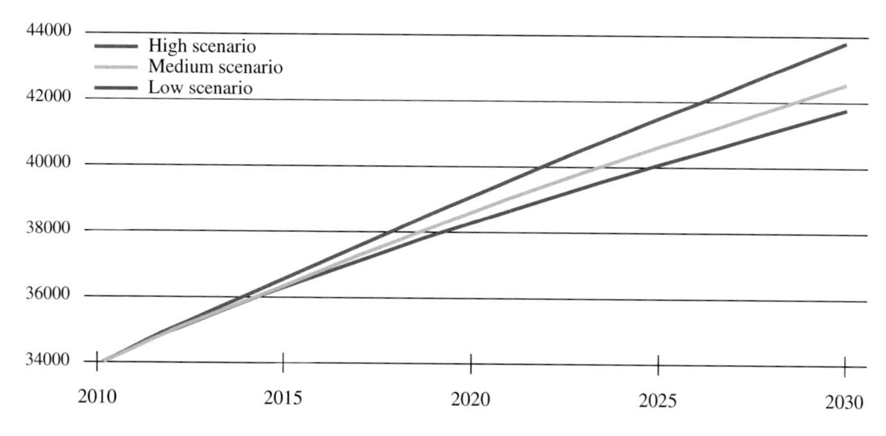

Fig. 3.2 Population evolution in the greater Yverdon-les-Bains urban region [redrawn from (SCRIS 2014)]

economic levels since Yverdon-les-Bains provides approx. 40 % of jobs in the Nord Vaudois region (Ville d'Yverdon-les-Bains 2009).

Furthermore, significant population growth is expected over the coming years since the arrival of some 6000 inhabitants and 3000 further jobs are forecast by 2020 (Fig. 3.2). This rapid expansion is related to both the socioeconomic development of the Yverdon-les-Bains conurbation itself and the overflow phenomenon of the Lake Geneva area which is struggling to respond to demand in terms of reasonably-priced housing and is characterised by a prolonged, even chronic shortage.

In this context, the Gare-Lac sector is of particular strategic significance since it is likely to host over 3000 inhabitants and 1500 jobs in a new, densely populated urban sector located directly in the geographic heart of the conurbation, in the area of influence of a major national railway station and, last but not least, with a beautiful lake-side landscape as a backdrop.

3.3 Localised Gare-Lac Master Plan

Today the Gare-Lac sector is looking for a fresh boost. With a mixture of buildings in search of new uses, abandoned lots and derelict factories, it resembles an area looking for an identity (Fig. 3.3). So, for Yverdon-les-Bains, it does not only represent strategic constructible potential in quantitative terms but it is also an opportunity to give a whole chunk of land a new identity and reclaim its urban lakeside.

This process was set in motion with the organisation of a Parallel Studies Competition (PSC) jointly involving the Municipality, the Region and the Canton. The aim was to define a development concept to rehabilitate the current urban

Fig. 3.3 View of the derelict railway lines at the eastern edge of the site. (*Source* LAST, EPFL; August 2012)

brownfield sites: a new part of town offering housing, office areas, shops and public spaces. The challenge was also to imagine original solutions to reduce the way the railway line cuts the area off from the rest of town, to create a pedestrian link-up along the Canal Oriental to connect the historic town centre with the lake and then study potential for a new harbour.

Following this procedure, which was carried out from September 2006 to June 2007, the interdisciplinary team led by Bauart Architects was declared the winner. This resulted in the drawing up of a Localised Master Plan (LMP), which is soon to become operative (Bauart et al. 2013).

Today, this Localised Master Plan (LMP) defines the required conditions and organization for the appearance of new time periods. It particularly highlights the enhancing of public areas associated with high urban density, as well as evolutive planning which will be capable of managing different elements of the operation over the given time span. The new neighbourhood therefore aims to become no less than a "flagship" for sustainable development In Yverdon-les-Bains (Bauart et al. 2013). Its general objectives include the ambition to reach the 2000-Watt society standard, to encourage soft mobility, to minimise the impact of individual motorized traffic (IMT) and encourage biodiversity. Figure 3.4 shows the major spatial features of the Localised Master Plan Gare-Lac.

Apart from regulatory aspects, which are beyond the scope of this report, the approach proposed depends on a precisely-defined understanding of the land and landscape. By locating the new sector to be rehabilitated in a larger urban ensemble,

Fig. 3.4 Localised Master Plan (LMP) of the Gare-Lac area [redrawn from (Bauart et al. 2013)]. *1* Old town; *2* Place d'Armes; *3* New town; *4* Parc des Rives; *5* Lake access

it recognizes the morphological importance of a series of structuring layers which extend successively from the historical town centre down to the lake (Fig. 3.4):

(1) The *Old town* and the first historical layers.
(2) The *Place d'Armes*, which plays the role of pivotal position between the old and the new town.
(3) The *New town*—densely populated, contemporary, urban and ecological—which hosts the new built sectors by integrating certain existing buildings and proposing new volumes. This sector will serve as the specific site for the project considered here, as a space having both a realistic and exploratory character.
(4) The *Parc des Rives*, an extensive wooded area whose specific layout will allow the creation of a landscaped unit involving a wide range of sports and cultural activities.
(5) The *Lake access*, which resembles a huge open meadow with a new sand surface providing a more attractive lake-side access.

References

Aubert E (1995) Histoire d'Yverdon I. Des temps préhistoriques à la conquête bernoise. Schaer, Yverdon-les-Bains

Bauart et al (2013) Plan Directeur Localisé Gare-Lac. Commune d'Yverdon-les-Bains, Yverdon-les-Bains

De Raemy D, Auderset P (1999) Histoire d'Yverdon II. De la Révolution vaudoise à nos jours. Schaer, Yverdon-les-Bains

Kasser G (1969) Yverdon: Eburodunum. Ur-Schweiz 33:54–57

Nast M (2006) Terres du lac. L'histoire de la correction des eaux du Jura. Gassmann, Nidau

SCRIS (2014) Perspectives de ménages 2010-2030. Demande de logements et population active. Canton de Vaud. Lausanne

Ville d'Yverdon-les-Bains (2009) Prix Wakker 2009. Patrimoine, Yverdon-les-Bains

Weidmann D (2014) Yverdon-les-Bains. In: Dictionnaire historique de la Suisse. Gilles Attinger, Hauterive

Chapter 4
Results

Abstract The chapter presents the results of the assessment of three scenarios developed in the framework of the case study of the Gare-Lac neighbourhood in Yverdon-les-Bains (Switzerland). The behavioural scenario clearly works out to be the most energy-efficient, whereas the technological scenario emerges as the most energy intensive for the three indicators considered (total primary energy, non-renewable primary energy and global warming potential). The symbiotic scenario falls more or less halfway between these two scenarios. In addition, it appears that none of the three scenarios manages to fully meet the intermediary objectives for 2035 defined by the vision of a 2000-Watt society. In reaction, a hybrid scenario is then developed to pragmatically combine technological innovations, residents' behavioural changes and the enhancement of potential resources by short circuits. This latter scenario proves that it is possible to reach most intermediary goals for 2035 while keeping to a holistic, sustainable and balanced approach which is not strictly conceived in terms of energy sobriety.

Keywords Sustainable neighbourhood · 2000-Watt Society · Energy consumption · Industrial ecology · Urban agriculture

Before presenting the results, we must point out that the values obtained include several approximations, bearing in mind the fact that estimations were required owing to the absence of precise data for certain parameters. However, the scenario approach remains a good way of integrating this variable degree of precision and of prioritizing a relative value comparison.

4.1 Local Resources

In order to adequately assess the energy autonomy of the future Gare-Lac neighbourhood, listed local resources are allocated according to a principle of territorial representativeness. If the resource or energy production installation is intended for the whole Yverdon-les-Bains conurbation, 7 % will be devoted to the neighbourhood power supply (i.e. the ratio between the neighbourhood population and the greater urban area).

S. Lufkin et al., *Strategies for Symbiotic Urban Neighbourhoods*,
SpringerBriefs in Applied Sciences and Technology,
DOI 10.1007/978-3-319-25610-8_4

For resources shared only with the Municipality, this figure becomes 14 %. For installations located In the immediate vicinity of the new neighbourhood, such as the thermal baths of Yverdon-les-Bains, this figure may even rise to approx. 30 % as it has been estimated that it was not realistic to consider that production be shared with the rest of the town because of thermal losses due to transport.

4.1.1 Thermal Waste

The thermal baths of Yverdon-les-Bains and the water purification plant (WPP) are considered as the two main candidates likely to contribute to local symbioses by thermal 'waste' recuperation. The thermal baths of Yverdon-les-Bains are located right in the centre of town (see Fig. 4.1). Currently water is extracted at 28° and gas-heated to 32°. Pools are regularly refreshed and at the present time surplus water is drained into Canal du Buron. Potential recuperation could reach approx. 320 thermo kW per annum. Given the proximity of this site to the case study perimeter (see above), it has been decided to allocate approx. 30 % of this energy to the new neighbourhood.

The WPP also presents a high potential for the use of waste heat. Currently 11,000 m^3 of water are purified every day. According to the Energy Department of Yverdon-les-Bains (SEY), this figure could reach 17,000 m^3 by 2035. By deducting the WPP's internal requirements in terms of electricity and heat as well as the 420 thermo MWh supplied to the neighbouring greenhouses, it was estimated that it would be possible to recuperate a power output of 3.2 thermo MW.

Moreover, waste heat producers near the case study perimeter have been listed (see Fig. 4.1). For the sake of realism, recuperation of their waste heat has not been taken into account in the scenarios. The SEY indeed believe that it is more interesting to use this process downstream from the WPP by applying the "Sewage Recovery System" approach.

This heat recovery process from wastewater was notably introduced in the False Creek neighbourhood in Vancouver (Canada) (FVB Energy Inc. 2006). By way of comparison, in Switzerland each year 6000 thermo GWh (i.e. 7 % of heating and domestic hot water requirements) are not made use of in wastewater. Wastewater in drainage systems varies between temperatures of 10 and 20 °C, thus offering a reliable basis for heat recovery, as well as for the generation of cold for air conditioning. Note that the temperature of wastewater is not sufficient to be used as is. A heat pump is necessary to elevate the temperature to harvestable level (50–70 °C). A heat exchanger provides another solution, which guarantees a separation between wastewater and domestic hot water. It can be installed at the bottom of the culvert or at the water purification plant (WPP).

Energy recover from wastewater is carried out using different systems: directly in the building, in the drainage system, up- or downstream from the WPP (Bühler 2008). The latter option will be prioritized in Yverdon-les-Bains.

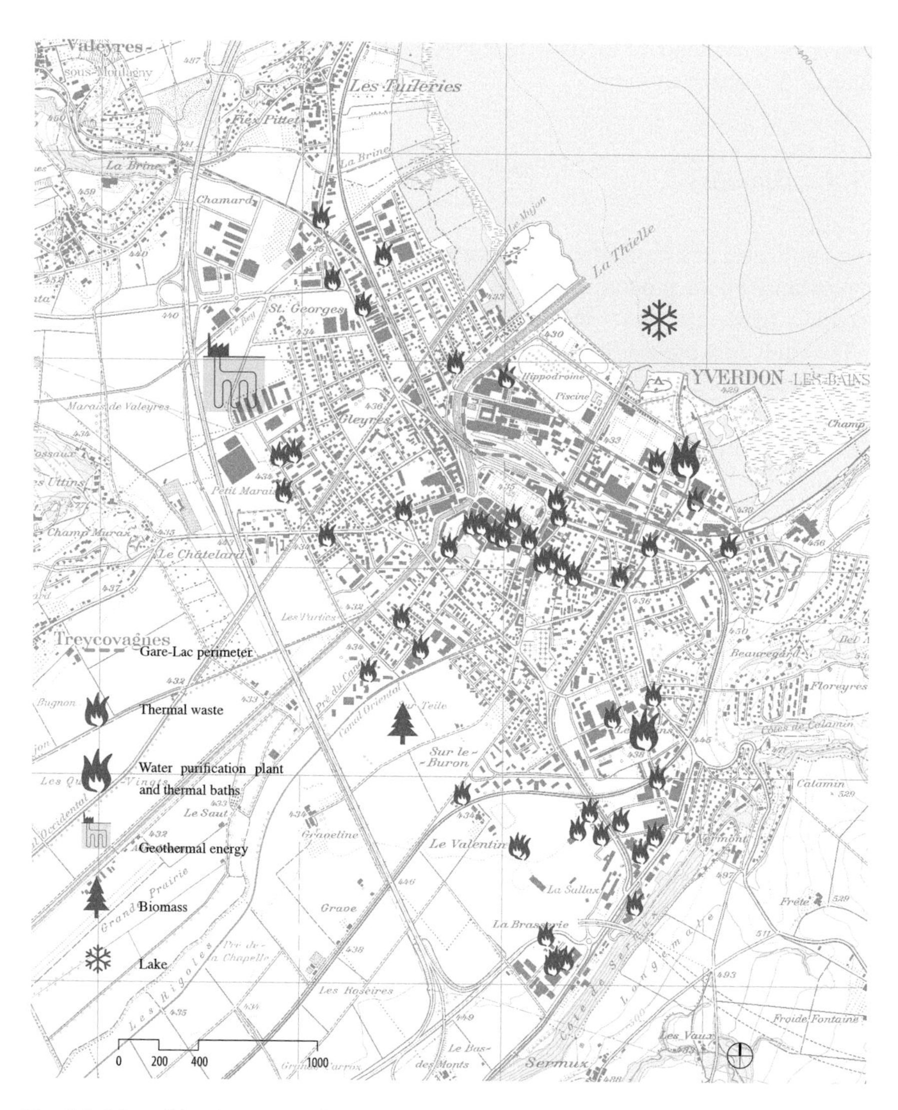

Fig. 4.1 Map of local resources for the Gare-Lac sector, Yverdon-les-Bains

4.1.2 Geothermal Energy

On the basis of information provided by the Municipality of Yverdon-les-Bains, a deep geothermal energy cogeneration project should be completed in 2017. Alone, It represents a supply of 5 MW of electricity and 60 thermo MW, which corresponds to the consumption of about 3200 households. According to hypotheses worked out in the framework of this research, 14 % will be attributable to the new Gare-Lac neighbourhood. This percentage corresponds to the ratio between the number of inhabitants in the new neighbourhood and the population of the commune of Yverdon-les-Bains.

4.1.3 Biomass

It was estimated that organic waste from the neighbourhood as well from as animals in proximity could produce up to 114 kW of electricity thanks to the heat and electricity from agricultural biogas (combined heat and power).

The study of a project for a wood-fired power station is underway. On the other hand, it should be pointed out that its geographic location is competing with the site planned for the geothermal energy plant. The potential production of wood-chips within a 50 km radius of the power station's site would be approx. 150,000 m^3 annually, which corresponds to 17 thermo MW. Here again, the hypothesis is to use 14 % of this capacity for the requirements of the future neighbourhood, i.e. 2.4 MW.

4.1.4 Solar Energy

The SEY have planned for annual growth in energy supply thanks to photovoltaic technology of 0.5 MW until 2035, solely for the area of the Commune, i.e. a total of 13 MW including annual capacity. As with biomass and geothermal energy, 14 % of this potential is allocated to the future neighbourhood.

4.1.5 Wind Power

Regarding wind power, the relevant scale is the Nord Vaudois region. Two projects in particular should be completed: the first is at the Mollendruz (30 MW), and the second at Cronay (9.5 MW). Only 7 % of this supply will be used, corresponding to 2.8 MW of electricity.

4.1.6 Lake of Neuchâtel

The Lake of Neuchâtel is the last energy source surveyed in the Yverdon-les-Bains area. The waters of the lake represent enormous potential as a source of heat and cold. The principle involves establishment of a hydrothermal network using and distributing lake water to cool the neighbourhood's buildings.

Pumped deepwater flows to buildings in a pipeline network linking the pumping station to the various buildings. In the summer, buildings can thus be cooled on the spot by an exchanger and by adapting the control system. In winter, inversely, high-efficiency heat pumps heat the buildings.

For environmental reasons, lake water exploitation is not desired by the Yverdon-les-Bains conurbation's cantonal and communal political authorities. So, in spite of this important potential, lake water will not be considered as a thermal resource in this research.

4.2 Performances of the Scenarios

4.2.1 Current Situation

Strictly speaking, this first stage is not a scenario. It concerns the present situation which gives an idea of the status of the neighbourhood if it were inhabited at full capacity. The number of inhabitants (3810) and jobs (1260) was specified by the Localised Master Plan. This present status serves as a point of reference which then enables data adaptation depending of hypotheses developed for each scenario. The energy-related behaviour and profile of the neighbourhood's residents are comparable with the present Swiss average (for reasons of data availability, these are more precisely figures for 2010).

Thus, neither supply for the neighbourhood nor its residents' lifestyles are defined in a prospective way and are more of an optimized modelling of the energy consumption and carbon footprint of contemporary Swiss society. Due to uncertainty regarding energy supply, we consider that the Swiss electricity mix remains unchanged.

Table 4.1 Estimated consumption for the Gare-Lac neighbourhood at full capacity (3810 residents and 1260 jobs), according to the three indicators chosen: total primary energy (W/pers), non-renewable primary energy and global warming potential GWP (kg eq. CO_2/year). The residents' energy-related profile corresponds to the Swiss average in 2010.

4.2.2 Technological Scenario

This first scenario relies on the use of cutting-edge technologies to reduce energy consumption and greenhouse gas emissions in the future neighbourhood. No in-depth change in user behaviour is expected and current trends will continue.

4.2.2.1 Energy Supply

The evaluation of the potential of the different local resources available worked out in the framework of this research (cf. Sect. 4.1) brought to light the important potential for geothermal heat cogeneration from the region's hot fractured rock. In the

Table 4.1 Provides an estimate of the energy consumption of the current situation for the three selected indicators

	Total primary energy (W/pers)	Non renewable primary energy (W/pers)	Global warming potential (kg eq. CO_2/year)
Buildings	1800	1693	1627
Construction	446	446	1086
Heating	230	212	91
Domestic hot water	212	195	84
Ventilation	153	141	62
Lighting/appliances	759	698	302
Mobility	1740	1642	3099
Car	1051	1026	1932
Plane	317	314	639
Train	115	51	47
Other public transport	32	31	62
Remaining	225	220	419
Infrastructures	601	556	666
Equipment	94	74	85
External installations	150	138	250
Remaining	357	344	331
Food	750	690	1710
Local and imported	750	690	1710
Goods and services	750	690	1012
Consumption	750	690	1012
Total	5641	5270	8110

technological scenario, 14 % of this energy supply is to be allocated to the future neighbourhood, i.e. 700 kW of electricity and 8.4 thermo MW. Wind-powered energy also contributes to the sector's energy supply with approx. 2.8 MW of electricity.

Inside the perimeter of the neighbourhood, solar power plants are set up on the large rooftop areas (see Fig. 4.2). The electricity produced (about 1.82 MW) will be stored in the form of hydrogen which will facilitate both the compensation of production intermittence and the supply of energy to part of the neighbourhood's individual vehicles (see the paragraph on mobility).

4.2.2.2 Buildings

Buildings in the technological scenario reach Minergie A® standard, the highest level of operating performance. This corresponds to 50 kWh/m^2 for construction, 15 kWh/m^2 for operating sub-categories: heating, DHW and ventilation, and

45 kWh/m^2 for lighting and domestic appliances. The large amount of insulation materials makes the construction method relatively massive which, in return, weighs down the construction energy burden. This is a new generation of "intelligent" buildings: lamps and electric appliances are highly efficient but the multiplication of electronic and home automation equipment tend to counterbalance these energy savings.

Systems for using renewable energy are integrated systematically in the buildings (photovoltaic panels, etc.). At the urban form level, this approach can be seen in the volumetric work to optimize the orientation of rooftops to maximize the solar panels' performance (Fig. 4.2).

In parallel, trends to reduce the size of households continue. They are accompanied by an increase in the surface area per inhabitant: this is 60 m^2 per person compared to 50 m^2 currently and 34 m^2 in 1980. This layout of large flats for small families has a significant impact on energy consumption.

4.2.2.3 Mobility

The daily distance covered by users is unchanged compared to current trends. On the other hand, technology developed for each type of transport is evolving. It is

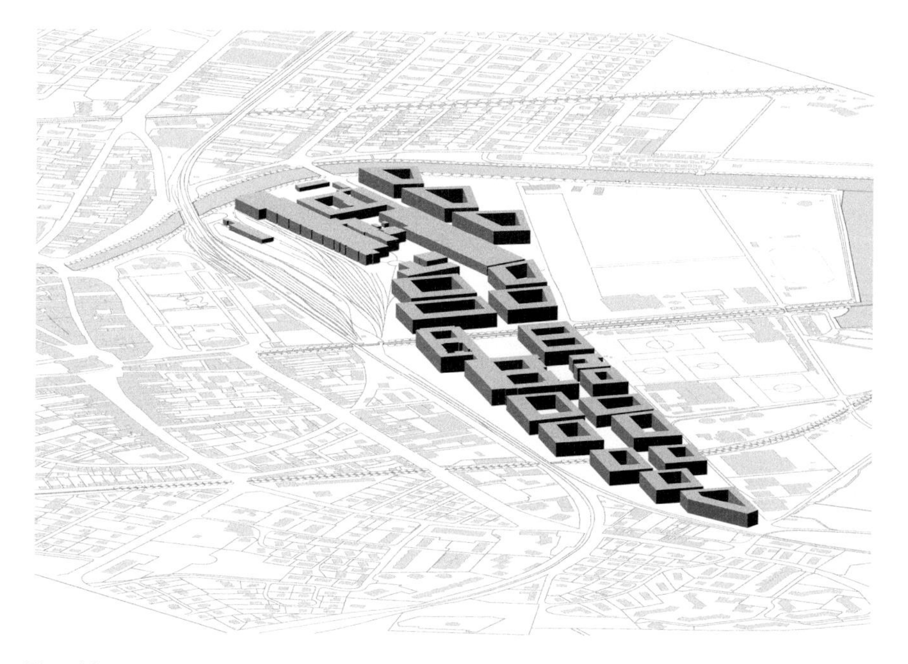

Fig. 4.2 Technological scenario. Visualisation of urban form. Rooftops are given specific volumetric attention with a view to installing optimized photovoltaic systems. Inhabitants: 3810, jobs: 1260, gross floor area—housing: 228600 m^2, gross floor area—offices: 31500 m^2, total gross floor area: 260100 m^2

suggested that these technologies, such as hydrogen-powered vehicles, will have made sufficient progress by 2035. 50 % of users will be using a hydrogen-powered vehicle (from renewable sources) for their individual motorised transport (IMT) movements, which represent 9957 person-kilometers per year (pkm/yr). The remaining 50 % will use a conventional car.

4.2.2.4 Infrastructures

The organisation of public spaces remains conventional and most surfaces are mineralized. Low-energy streetlights provide all the neighbourhood's public lighting.

4.2.2.5 Food

No major change for this category compared to the current situation. The present percentage of vegetarians in Switzerland remains stable but a small portion of the future neighbourhood chooses locally produced organic food and fresh seasonal products. The average energy consumption for food falls 4 % compared to the current situation.

4.2.2.6 Goods and Services

The emergence of high-tech products as a social marker continues and increases. As in the buildings category, global increase in consumption (multiplication of electronic appliances) tends to compensate the improvement in energy performance, in spite of widespread distribution of information concerning the consumption of fossil carbon (carbon labels and personal carbon calculators). The resulting figure of 750 W for total primary energy consumption per person, based on the estimation of the current situation, therefore remains unchanged for the technological scenario.

Table 4.2 Technological scenario. Assessment of energy-related performance according to the three indicators chosen for this research: total primary energy (W/pers), non-renewable primary energy (W/pers) and global warming potential (GWP) (kg eq. CO_2/year).

4.2.3 Behavioural Scenario

This second scenario takes the opposite stance to the technological scenario, by focusing in priority on the hypothesis of a significant evolution in user behaviour:

Table 4.2 Provides an estimate of the energy performances of the technological scenario for the three selected indicators

	Total primary energy (W/pers)	Non renewable primary energy (W/pers)	Global warming potential (kg eq. CO_2/year)
Buildings	642	428	914
Construction	401	401	867
Heating	31	7	7
Domestic hot water	25	5	6
Ventilation	49	4	9
Lighting/appliances	137	11	25
Mobility	2299	1199	2238
Car	1610	583	1076
Plane	317	314	639
Train	115	51	47
Other public transport	32	31	62
Remaining	225	220	414
Infrastructures	551	491	603
Equipment	44	9	21
External installations	150	138	250
Remaining	357	344	331
Alimentation	720	663	1638
Local and imported	720	663	1638
Goods and services	750	690	1012
Consumption	750	690	1012
Total	4962	3471	6405

frugal consumption, intentional simplicity, decrease in consumerism and appreciation of a slower lifestyle. In this scenario, the driving force behind energy transition is the decrease in demand thanks to a change in certain current social practices.

4.2.3.1 Energy Supply

Two energy supply sources were focused on for the behavioural scenario: cogeneration by agricultural biogas (from liquid hog manure), and a wood-fired power station. Study of the latter project, on the Freymond site, is currently ongoing. 14 % of these resources are allocated to the new neighbourhood, i.e. a total of 2.4 thermo MW and 0.1 MW of electricity.

4.2.3.2 Buildings

To minimize grey energy, buildings in the new neighbourhood are light constructions made from local wood (SIA 2004). Indeed, wood has many environmental advantages particularly because its production requires less fossil energy than other construction materials (Mahapatra and Gustavsson 2009) and that the carbon in wood is sequestered rather than emitted into the atmosphere. Moreover, local provenance limits energy used for transport. The Minergie standard alone is demanded.

The resident becomes an actor in her habitat; she takes part in some of the building work and maintenance. In addition, in order to save on costs and energy, she accepts lower heating temperatures at home. Some areas (living rooms, kitchens, bathrooms, etc.) and equipment (wifi, freezers, etc.) are shared which results in an increase in the size of households. Thanks to the sharing of living spaces and the collective habitat, but with the same comfort standards, the surface area per inhabitant thus drops to 40 m^2 in the behavioural scenario. In comparison, this figure is about the equivalent to the average in 1990. This reduction gives a total decrease of 28 % in the necessary surface area required to accommodate the residents of the neighbourhood.

4.2.3.3 Mobility

Generally speaking, kilometres travelled drop; local commitment and tourism are favoured in comparison with travel to exotic destinations. Some plane journeys are also replaced by the train. This effort is definitely worth it since short flights consume the most energy per person per kilometre (KBOB 2012). This is notably due to the fact that taking off and landing are the greediest in kerosene. Within the new neighbourhood, distances of less than 1.5 km covered by individual motorised transport (IMT) are replaced by soft mobility. This short distance corresponds to a journey time of approx. 20 min on foot of 5 min by bike.

The search for economic and supportive solutions between residents leads 30 % of them to choose car-sharing (with cooperatives like Mobility), thus reducing the kilometres travelled in IMT and, in return, there is an increase in km travelled on public transport. Overall, this transfer is advantageous because of the latters' better primary energy factor. Moreover, for strictly work-related travel, 40 % of users favour car-pooling which raises average car occupation from 1.12 (OFS 2012) today to 2 persons per vehicle. The number of cars per household is thus lower, with some people choosing not to have a car at all and others giving up their second car. Sometimes people also choose a return to animal-drawn vehicles.

4.2.3.4 Infrastructures

Large collective areas freed up by the reduction of built-up density (see paragraph on Buildings) are devoted to urban agriculture (see Fig. 4.3). The organization of public spaces is more economical like the unmown grass borders along roads which increase biodiversity in the neighbourhood as well as reducing maintenance work. In much the same way, a decrease in the number of kilometres covered and the increase in soft mobility allow use of road surfaces which reduce environmental impact.

4.2.3.5 Food

We see significant modifications in this category compared with the current situation. Food-related consumption depends very much on the residents' behaviour; unlike other categories such as buildings or infrastructures where the technology installed or the energy supply source are the major parameters influencing energy consumption.

In the behavioural scenario, 70 % of inhabitants have a vegetarian, local, seasonal and organic diet (compared with some 10 % currently). Extensive communal nursery gardens/allotments (see Fig. 4.3) are available for residents to encourage short supply chains, local sale and exchange of fruit and vegetables. Waste and consumption of refined products such as chocolate, coffee and alcohol are also avoided. Special attention is paid to waste reduction and minimization of food packaging. The combination of these different efforts enables users in this scenario to reduce their individual consumption by 46 % compared to the average Swiss household.

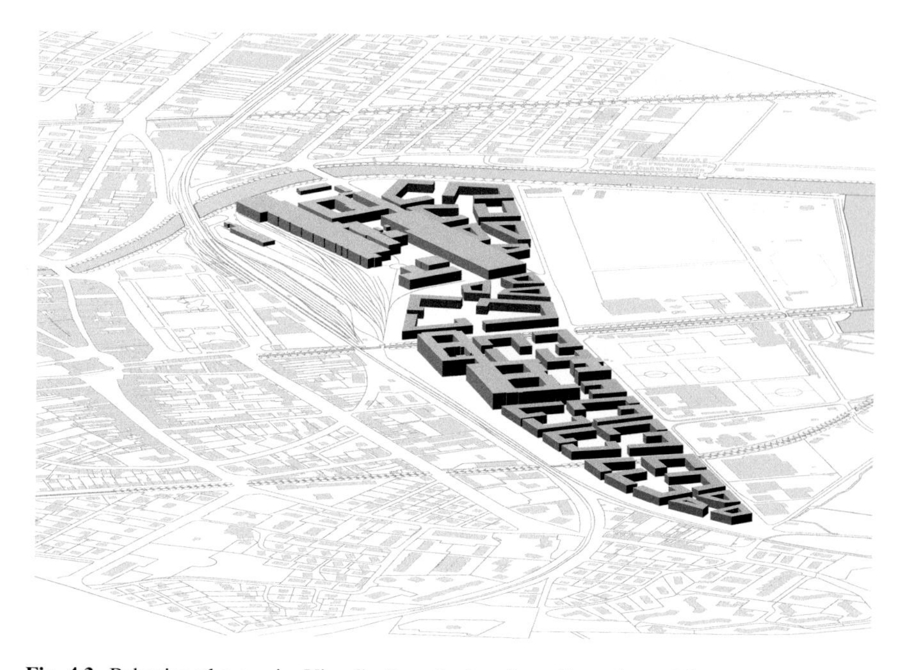

Fig. 4.3 Behavioural scenario. Visualisation of urban form. Extensive public spaces dedicated to urban agriculture also provide popular gathering places for neighbourhood residents. Inhabitants: 3810, jobs: 1260, gross floor area—housing: 152400 m^2, gross floor area—offices: 6300 m^2, gross floor area—activities: 50400 m^2, total gross floor area: 209100 m^2

Table 4.3 Provides an estimate of the energy performances of the behavioural scenario for the three selected indicators

	Total primary energy (W/pers)	Non renewable primary energy (W/pers)	Global warming potential (kg eq. CO_2/year)
Buildings	651	449	400
Construction	215	78	172
Heating	7	6	17
Domestic hot water	8	7	20
Ventilation	131	112	60
Lighting/appliances	288	246	131
Mobility	1483	1353	2515
Car	847	827	1556
Plane	186	185	387
Train	184	81	77
Other public transport	41	41	81
Remaining	225	220	414
Infrastructures	508	479	578
Equipment	37	31	30
External installations	150	138	250
Remaining	321	310	298
Food	406	373	923
Local and imported	406	373	923
Goods and services	638	587	860
Consumption	638	587	860
Total	3685	3240	5275

4.2.3.6 Goods and Services

In this area we are looking at just the opposite effect to the technological scenario. The symbolic depreciation of consumer goods implements self-production of furniture and clothes, recovery and diversion. The responsible user naturally looks to sustainable, repairable objects. Health-linked consumption also drops thanks to 'alternative' techniques such as natural medicines, healthier nutrition, and home care for the elderly and children. It is estimated that these strategies will bring about a decrease of some 15 % in energy consumption for the average Swiss household.

Table 4.3 Behavioural scenario. Assessment of energy-related performance according to the three indicators chosen for this research: total primary energy (W/pers), non-renewable primary energy (W/pers) and global warming potential (GWP) (kg eq. CO_2/year).

4.2.4 *Symbiotic Scenario*

This scenario greatly enhances the possibilities for urban and industrial symbioses in reducing the neighbourhood's impact on the environment. The approach implies optimal use of energy and material exchange networks at every level. Regarding behavioural issues, this scenario falls halfway between the two previous ones, with users acting according to network and partnership logics.

4.2.4.1 Energy Supply

The two main sources exploited for the symbiotic neighbourhood's energy supply are the thermal baths of Yverdon-les-Bains and, above all, the water purification plant, which represents a high potential for recovery of waste heat (see Sect. 4.1). The combination of these two resources allows for the generation of 3.4 thermo MW and 0.1 MW of electricity.

4.2.4.2 Buildings

The symbiotic scenario demands the Minergie-P standard for the new constructions, i.e. 90 MJ/m^2 (25 kWh/m^2). Moreover, recycled construction materials are given preference. Different strategies are set up to transform lost waste into resources. For individual buildings, heat waste is recycled, especially by wastewater and ventilation heat recovery.

The average surface area per resident has been fixed at 50 m^2, i.e. a stabilisation compared to the current situation. In order to enhance exchanges at neighbourhood level, this scenario furthermore stands out for its significant functional diversity. In addition to surface areas devoted to craftsmanship, certain non-polluting industrial activities are also encouraged to move to the sector.

The urban form proposed by the symbiotic scenario (Fig. 4.4) is characterized by megastructures hosting a combination of functions which may at first glance appear rather incompatible, such as one of the buildings which houses the new bus depot of the City of Yverdon-les-Bains on the ground floor and businesses, shops and residential units upstairs, or vertical farms (see section on Food).

4.2.4.3 Mobility

Consumers feel responsible and act collectively as their attitude to mobility evolves. They tend to choose smaller, lighter vehicles which consume less and even sometimes run on biodiesel (50 % of users). The public transport network has substantially expanded. It is much more attractive to users who use collective forms of mobility more often. Over long distances, across Europe, for example, rail

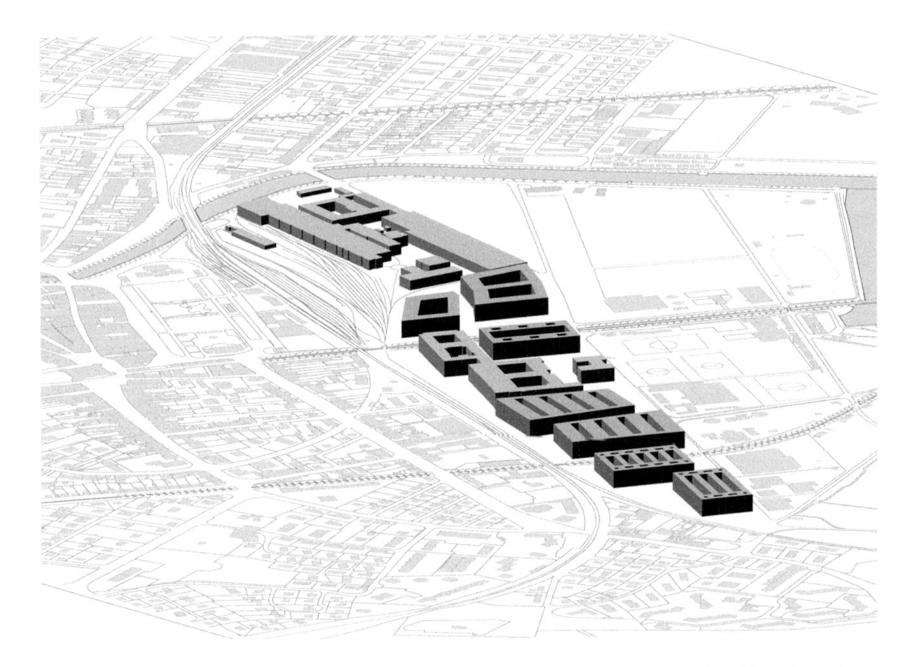

Fig. 4.4 Symbiotic scenario. Visualisation of urban form. Large multifunctional blocks enhance short circuit exchanges at the building level, as well as on the neighbourhood scale. Heat waste from the surrounding districts is exploited as an energy source. This includes heat (metal construction businesses, wastewater from the water purification plant) and cold sources (lake water). Inhabitants: 3810, jobs: 1260, gross floor area—housing: 190500 m^2, gross floor area—offices: 6300 m^2, gross floor area—activities: 25200 m^2, gross floor area—industries: 40320 m^2, total gross floor area: 262320 m^2

transport is preferred to air travel for leisure trips. A more detailed description of the figures for different means of transport is to be found in the annexes.

4.2.4.4 Infrastructures

The strong enhancement of the collective dimension can be seen in the organization of large, convivial public spaces. These public spaces are set out as permeable areas which will favourably influence the management of water within the neighbourhood and of biodiversity in general. Due to the difficulty of quantifying the impact of this organization, current figures have been used for reference in this category.

4.2.4.5 Food

Here again, the symbiotic scenario falls halfway between the two previous scenarios. Neighbourhood residents maintain a balanced diet we can describe as "healthy and environmentally friendly". Hypotheses point to the stabilization of the current percentage of vegetarians, at some 10 %. On the other hand, 70 % of users eat local, seasonal food, reduce meat products and avoid excess. Vertical farms allowing for the optimization of waste "grey" water and compost from household organic waste as natural fertilizers are set up in the neighbourhood megastructures to accompany this change in lifestyle.

50 % of consumers are mindful of waste and 20 % of them give up refined products. On average, these measures lead to a reduction of approx. 13 % in energy consumption.

4.2.4.6 Goods and Services

There is no radical transformation in the consumption of goods and services in the symbiotic scenario. Some users favour repairable or recyclable products, as well as eco- or regenerative designs. These trend modifications give a final result of approx. 10 % less than the current situation.

Table 4.4 Symbiotic scenario. Assessment of energy-related performance according to the three indicators chosen for this research: total primary energy (W/pers), non-renewable primary energy (W/pers) and global warming potential (GWP) (kg eq. CO_2/year).

4.2.5 Comparison of the Scenarios

The comparison of the energy performances of the scenarios (Figs. 4.5, 4.6 and 4.7) soon reveals that all three are more effective than the current situation. Furthermore, ranking the scenarios is almost immediate: the behavioural scenario clearly works out to be the most energy-efficient, whereas the technological scenario emerges as the most energy intensive for the three indicators considered (total primary energy, non-renewable primary energy and global warming potential). The symbiotic scenario falls more or less halfway between these two scenarios, with the exception of non-renewable primary energy which marks it as being slightly less efficient than the two other scenarios.

It is important however to qualify these results. Indeed, only strictly energy-related aspects were considered in the evaluation. Further criteria linked to social (acceptability), economical (cost of technologies implemented) and environmental (global environmental impact) aspects would be necessary to complete assessment of the scenarios. For example, the radical vision embodied in the behavioural scenario is unrealistic as it represents a burden on the resident's individual freedom. This is a good illustration of the pertinence of holistic approaches which are at the core of the concept of sustainable development. Therefore, by only bearing these three indicators in mind it is difficult to pick out one of the scenarios as the most efficient globally.

Table 4.4 Provides an estimate of the energy performances of the symbiotic scenario for the three selected indicators

	Total primary energy (W/pers)	Non renewable primary energy (W/pers)	Global warming potential (kg eq. CO_2/year)
Buildings	778	690	731
Construction	185	185	451
Heating	89	76	45
Domestic hot water	98	83	50
Ventilation	127	108	58
Lighting/appliances	279	237	127
Mobility	1307	1187	2015
Car	676	643	1007
Plane	236	233	474
Train	138	61	58
Other public transport	32	31	62
Remaining	225	220	414
Infrastructures	543	513	611
Equipment	86	73	52
External installations	150	138	250
Remaining	357	344	331
Food	653	600	1485
Local and imported	653	600	1485
Goods and services	675	621	911
Consumption	675	621	911
Total	4005	3654	5775

The calculation of the three indicators carried out as part of this research (cf. Sect. 4.2) remains relevant because it allows us to highlight certain phenomena on a finer scale and requires close consideration of the results by consumption category.

Regarding the buildings, results are counterintuitive: the best-graded ones (from the behavioural scenario) are those which meet the least stringent standard (Minergie, whereas the technological scenario demands Minergie A). Two deciding factors explain these results. On the one hand, the floor area per resident: reducing this, in the behavioural scenario, to 40 m^2 per person (as compared to 50 m^2 currently) indeed leads to a decrease of approx. 30 % of the surface area necessary to host the neighbourhood's residents. On the other, the energy required for construction of the buildings, in the behavioural scenario, is reduced thanks to the construction method using local wood. Conversely, the heavy constructions involved in the technology scenario have a significant impact on the buildings' grey energy.

Regarding mobility, it turns out that this plays a major role in the three scenarios. In the technological scenario, for example, mobility represents half the total energy consumption. Its impact shrinks considerably for non-renewable energy, thanks to the use of hydrogen-powered vehicles. In the behavioural scenario, although mobility significantly drops, the use of conventional cars impacts heavily. The symbiotic scenario offers the most efficient alternative by combining use of public transport and biodiesel- and electricity-powered cars (power produced from renewable sources).

Finally, regarding food, findings are similar to mobility issues. It indeed turns out that, in the behavioural scenario, the biggest reductions concern meat consumption. However, results reveal that the impact of this change in eating behaviour on the total energy balance remains low. From a purely energy-related point of view, the effort made represents merely a marginal benefit, whereas reducing meat consumption requires an extremely important commitment on the part of residents.

Comparing the results obtained within this research with the intermediary objectives for 2035, as defined by the vision of a 2000-Watt society, it appears that none of the three scenarios manages to fully meet the targets. Unlike the technological scenario, total primary energy consumption in the behavioural and symbiotic scenarios is significantly less than the threshold of 4400 W/pers. Regarding non-renewable primary energy, results slightly exceed the target of 3300 W/pers. However, bearing in mind the uncertainties affecting some of the data, it may be considered that this order of magnitude is roughly comparable.

By contrast, the CO_2 emissions of the three scenarios are clearly superior to the intermediary objective of 3.2 tonnes. These results can be explained partly by the relatively pessimistic hypotheses used for the flow analyses. Due to the difficulty in obtaining reliable figures for 2035, the conversion factors used reflect the current state of technology. However, in the years to come, technological innovations, especially in the auto sector, should focus mainly on yield improvement. Proportionally, greenhouse gas should hence drop more than total fuel consumption. But forecasting this evolution is extremely complex and would require an in-depth analysis which is beyond the scope of the present study.

Comparison with targets for 2100 provides a good assessment of what still needs to be done.

It turns out from this comparison that none of the strategies, considered singly, is sufficient to globally “tackle” the energy-related transition targets alone. The behavioural scenario, which appears at first glance to be the most promising candidate, raises a whole series of issues at its acceptability level. Consumption in the technological scenario, beyond the unsolved economic feasibility issues raised, is still too high. Idem for the symbiotic scenario which would also require particularly intensive political support and investment in education before perspectives for urban symbioses can materialise.

We will therefore have to combine these strategies. This is precisely the goal sought in the last research phase which aims to develop this compromise approach with the formulation of an integrated or optimized scenario that pragmatically associates technological innovations, residents’ behavioural changes and the enhancement of potential resources by short circuits.

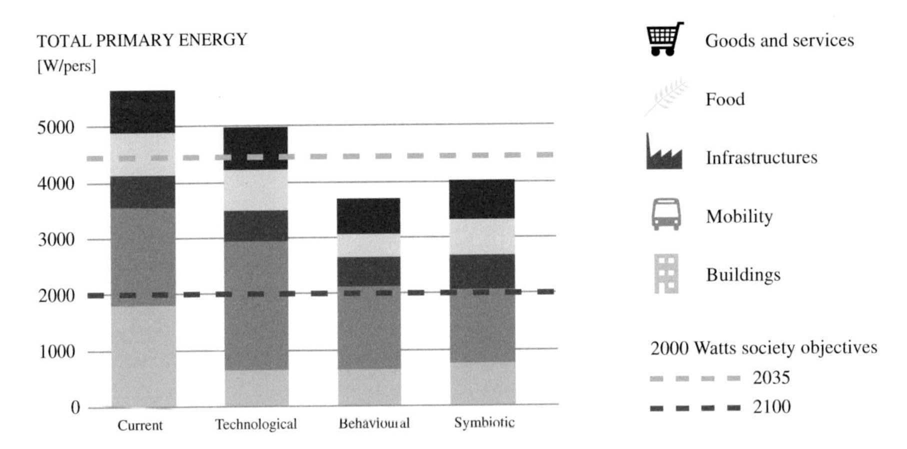

Fig. 4.5 Comparative results between the current situation and the technological, behavioural and symbiotic scenarios for total primary energy in (W/pers). Comparison with intermediary results for the 2000-Watt society for 2035 and 2100

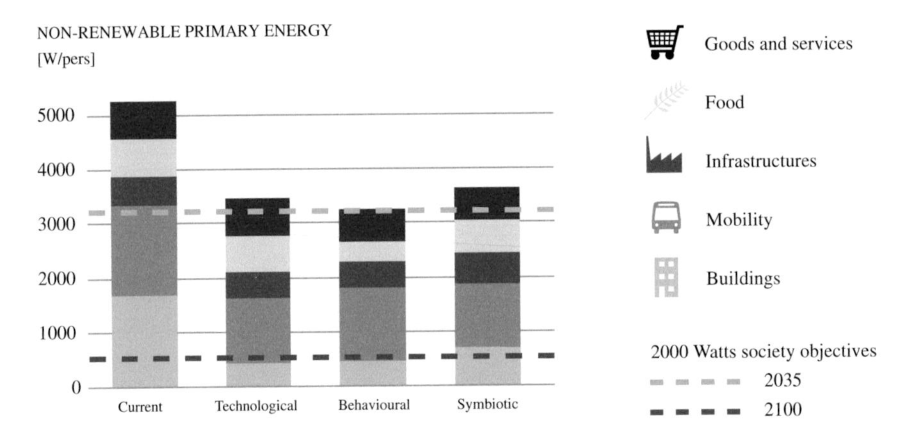

Fig. 4.6 Comparative results between the current situation and the technological, behavioural and symbiotic scenarios for non-renewable primary energy in (W/pers). Comparison with intermediary results for the 2000-Watt society for 2035 and 2100

4.3 Optimisation and Integration

The hybrid scenario is defined as a blending of the three previous scenarios. It assimilates the most effective strategies in order to obtain the highest global score both on the basis of the three energy-related indicators (Figs. 4.8, 4.9 and 4.10) and by integrating considerations linked with the economic (costs) and social (acceptability) aspects. As the result of a holistic approach, this scenario is therefore an

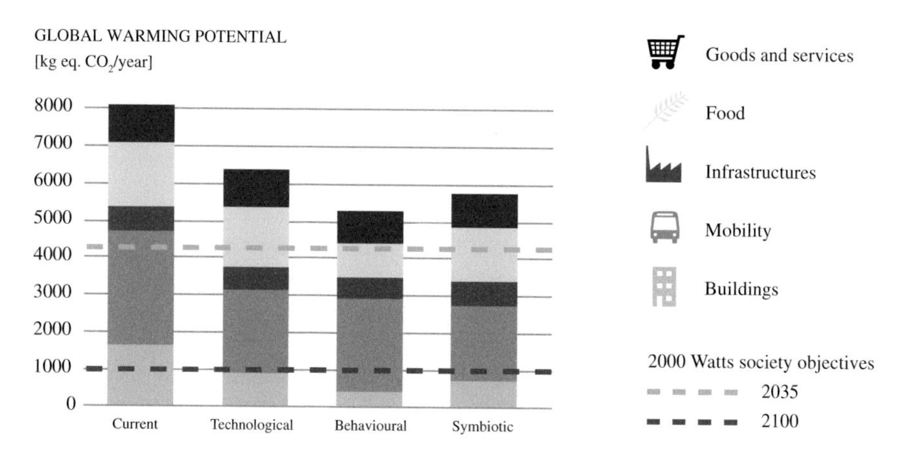

Fig. 4.7 Comparative results between the current situation and the technological, behavioural and symbiotic scenarios for global warming potential in (kg eq. CO_2/pers). Comparison with intermediary results for the 2000-Watt society for 2035 and 2100

attempt to reach a multi-criteria optimisation which fits in with the perspective of sustainable development.

The aim is also to test to what extent implemented technologies are complementary or exclusive. Indeed, the simultaneous use of two impact-reducing strategies does not necessarily result in the sum of their reductions.

Average user behaviour is comparable with the one in the symbiotic neighbourhood, i.e. more or less halfway between the technological and the behavioural

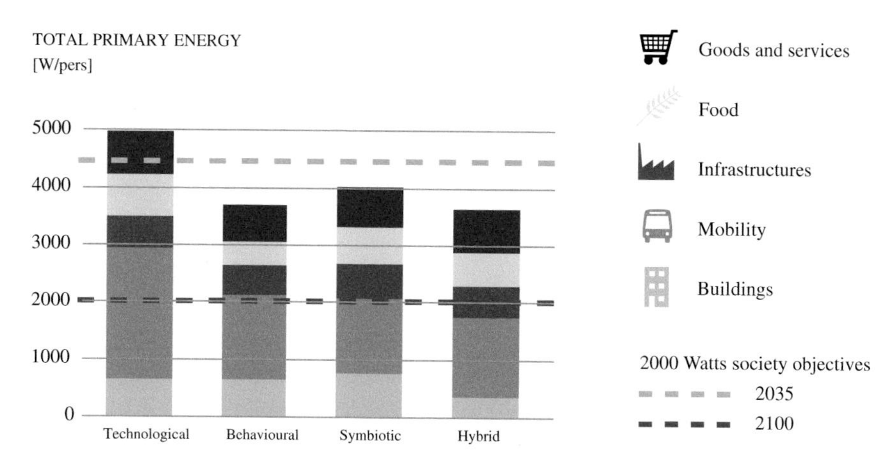

Fig. 4.8 Comparative results between the current situation and the technological, behavioural, symbiotic and hybrid scenarios for total primary energy in (W/pers). Comparison with intermediary results for the 2000-Watt society for 2035 and 2100

scenarios. This decision is coherent with the efforts to reach an equilibrium on which this integrated scenario is based.

4.3.1 Energy Supply

For this scenario, unlike what has been done previously, the strategy aims to start by quantifying needs in order to respond subsequently with the adequate energy supply. So as to encourage energetic self-reliance and optimize use of local resources, nearness of supply will be given priority.

Rooftops will be equipped by solar panels. Surplus electricity consumption is covered by a mix between wind power, geothermal energy, the installation of power-heat coupling (PHC) at the water purification plant (WPP) and photovoltaic panels at the communal level. Regarding heat consumption, priority for the thermal mix is given to waste heat from the WPP as well as the thermal baths. When necessary, geothermal energy helps compensate the shortfalls.

4.3.2 Buildings

This area is characterized by the implementation of hybrid solutions. For operation purposes, the Minergie-A standard has been assigned. For the heating, domestic running water and ventilation categories, the final consumption per surface area unit is estimated at 10 kWh/m^2, which is entirely possible with a consistent photovoltaic energy supply. On the other hand, due to the heavy construction method used, the traditional Minergie-A standard is very onerous in terms of non-renewable primary energy for construction. This is why priority is given to construction similar to the symbiotic scenario (Minergie-P), i.e. 90 MJ/m^2 (25 kWh/m^2).

Buildings are light constructions in local wood with a maximum of recycled materials and high thermal inertia. The surface area allocated per person is 50 m^2. Regarding energy supply, heat will be geothermal energy-powered and electricity from solar panels as well as, if necessary, from biogas cogeneration.

4.3.3 Mobility

In the integrated scenario, in the same way as in the symbiotic scenario, residents' mobility habits become more ethical. This can be seen more particularly in the decrease in kilometres travelled in individual motorised transport (IMT), thanks to car sharing (30 % of them) and carpooling (40 %). Preferred energy carriers for IMTs are hydrogen (15 %) and biodiesel (45 %). The remaining users use conventional fuel.

Leisure-related air travel in Europe is replaced by rail. Soft mobility is used for distances shorter than 1.5 km.

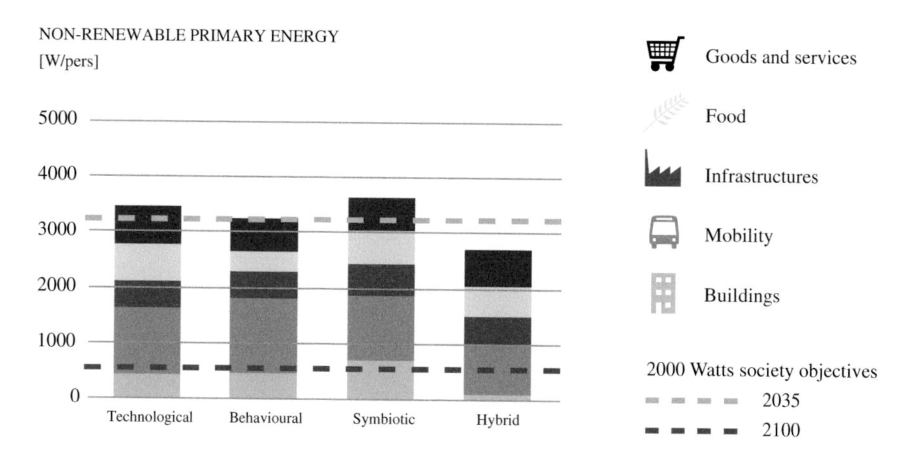

Fig. 4.9 Comparative results between the current situation and the technological, behavioural, symbiotic and hybrid scenarios for non-renewable primary energy in (W/pers). Comparison with intermediary results for the 2000-Watt society for 2035 and 2100

Fig. 4.10 Comparative results between the current situation and the technological, behavioural, symbiotic and hybrid scenarios for global warming potential in (kg eq. CO_2/pers). Comparison with intermediary results for the 2000-Watt society for 2035 and 2100

4.3.4 Infrastructures

As in the technological scenario, low-power street lighting is installed. Infrastructure-related consumption is thus slightly down on the current situation. Heat consumption is essentially supplied by waste heat from the WPP and by geothermal energy if top-up is necessary. Electric consumption Is supplied by a mix between the PHC from the WPP, solar energy, wind or geothermal power.

4.3.5 Food

As in the other areas, research aimed for an optimum result between the efforts made and the efficiency of the measures considered. For food, this especially implies a reduction of the energy-related outlay while maintaining a high degree of acceptability. 30 % of users are considered as vegetarians and half of these choose regional, seasonal, organic products and avoid excess. 30 % avoid waste but only 10 % will go without refined products. Thanks to these measures, energy reduction compared to the current situation is 22 %.

4.3.6 Goods and Services

Due to the low impact of user behavioural changes in the area of goods and services on total consumption, and in order to maximize the acceptability of this scenario, it

Table 4.5 Provides an estimate of the energy performances of the hybrid scenario for the three selected indicators

	Total primary energy (W/pers)	Non renewable primary energy (W/pers)	Global warming potential (kg eq. CO_2/year)
Buildings	359	111	261
Construction	197	71	158
Heating	17	4	4
Domestic hot water	14	3	3
Ventilation	26	7	19
Lighting/appliances	105	26	77
Mobility	1390	877	1722
Car	710	305	679
Plane	236	233	474
Train	178	78	74
Other public transport	41	41	81
Remaining	225	220	414
Infrastructures	560	511	582
Equipment	150	138	250
External installations	357	344	331
Remaining	357	344	331
Food	585	538	1331
Local and imported	585	538	1331
Goods and services	750	690	1012
Consumption	750	690	1012
Total	3645	2728	4908

was considered that the reference value provided by the evaluation of the current situation would remain identical for the integrated scenario.

Table 4.5 Hybrid scenario. Assessment of energy-related performance according to the three indicators chosen for this research: total primary energy (W/pers), non-renewable primary energy (W/pers) and global warming potential (GWP) (kg eq. CO_2/year).

References

Bühler T (2008) Récupération de la chaleur des eaux usées. Retour d'expériences faites en Suisse et en Allemagne. gwa 5:379–384

FVB Energy Inc. (2006) Potential heat sources for a neighbourhood energy utility, city of Vancouver, False Creek Precinct

KBOB (2012) Données des écobilans dans la construction 2009/1. Bern

Mahapatra K, Gustavsson L (2009) General conditions for construction of multi-storey wooden buildings in Western Europe, Östersund

OFS (2012) La mobilité en Suisse - Résultats du microrecensement mobilité et transports 2010. Office fédéral de la statistique (OFS), Neuchâtel

SIA (2004) SNARC. Méthode pour l'évaluation de l'écologie dans les projets d'architecture. Swiss society of engineers and architects (SIA), Zurich.

Chapter 5
Conclusions

Abstract The chapter presents the conclusions of the "Symbiotic Neighbourhoods" research project. In general, two main levers for action can be identified: mobility (type of vehicle and international mobility) and the heating needs per resident (strongly influenced by the energy-related reference surface area per resident). These results emphasize that, from the energy point of view, the urban densification issue is not so much a matter of built-up density but rather of the increased population density. The research also emphasises the low potential of strategies embodying a strict ideological vision of sustainable development: far from being contradictory, the technological, behavioural and symbiotic visions are complementary. The hybrid scenario shows that the pragmatic blending of these different strategies allows for the formulation of a concept that is relevant and adapted to the local context.

Keywords Sustainable neighbourhood · 2000-Watt society · Energy consumption · Industrial ecology · Urban agriculture

At the end of this research work, the time has come to draw some conclusions from the results obtained.

First of all, the results highlight how access to information has become a crucial issue. Very often stakeholders—individuals or public collectivities—try to assume their responsibilities regarding sustainable development issues but come up against an absence of reliable quantitative information which sometimes prevents them from making an informed choice. Moreover, debate regularly gets bogged down in sterile quarrels between partisans of technology and enthusiasts of a more sober lifestyle. It is therefore essential to be able to make a well-considered decision by identifying the main levers for potential action and priority strategies.

For example, if certain lifestyle-related aspects do indeed reduce energy consumption considerably, others only have marginal effects. In order to identify the "good" practices, it is thus ultimately essential to put into perspective the efforts to be made regarding the real effectiveness in the final balance. The best decisions will be made via this good understanding of information in view of providing a response to the rapid changes in energy policies expected over the coming decades.

S. Lufkin et al., *Strategies for Symbiotic Urban Neighbourhoods*,
SpringerBriefs in Applied Sciences and Technology,
DOI 10.1007/978-3-319-25610-8_5

Research also emphasises the low potential of strategies embodying a strict ideological vision of sustainable development. This type of approach is indeed based on pre-conceived ideas of sustainability and turns out to be totally incoherent when we attempt to apply it to part of a city. For example, although it remains unrealistic, use of the hydrogen-powered car seriously jeopardises the total primary energy balance of the technological neighbourhood and tends to cancel out all the efforts undertaken in the other categories. In the framework of this study, in order to build scenarios referring to as radical an ideological position as possible, this choice is nevertheless legitimate because it reflects an extreme high-tech trend in the area of mobility.

However, there is no miracle solution; far from being contradictory, the technological, behavioural and symbiotic visions are complementary. The hybrid scenario shows that the pragmatic blending of these different strategies allows for the formulation of a concept that is relevant and adapted to the local context.

At this stage of research and bearing in mind all the compulsory precautions resulting from the uncertainties of certain data, two main levers for action can be highlighted: mobility (type of vehicle and international mobility) and the heating needs per resident (strongly influenced by the energy-related reference surface area per resident). These results link up with those of previous studies and emphasize that, from the energy point of view, the urban densification issue is not so much a matter of built-up density but rather of the increased population density.

However, it remains true that the results of the flow analyses are still unsatisfactory for certain categories, notably for food, and above all for the consumption of goods and services. With a view to an extension of this research study, it would appear that the 'incomes' criteria would be a key issue for reflection on solving these uncertainties (Emelianoff et al. 2012).

Finally, comparing the results obtained with the goals of a 2000-Watt society in 2035 reveals the energy-related impacts of planning on the neighbourhood scale. The hybrid scenario proves that it is possible to reach the intermediary goals for 2035—with the exception of greenhouse gas emissions—while keeping to a holistic, sustainable and balanced approach which is not strictly conceived in terms of energy sobriety.

This would suggest that to achieve all the goals of the 2000-Watt society, in particular those linked with greenhouse gas, reflection which would merely envisage action at the neighbourhood level would turn out to be insufficient. And for a good reason, the concept of a 2000-Watt society can only be globally achieved by a general transition within the whole of Swiss society, particularly in the economy sectors, towards practices and technologies with a lower environmental impact.

Moreover, this study once again emphasizes the relevance of working at the neighbourhood scale with a view to approaching issues linked with energy and the sustainability of the built environment. The energy-related performance of the symbiotic and hybrid scenarios, without demanding draconian restrictions by the resident, indeed show the soundness of a reflection which goes beyond the individual building level in order to enhance exchanges between buildings. The results presented in this report also highlight the interest in pursuing these reflections which contrast approaches emerging from architecture, urbanism and industrial ecology at the neighbourhood level. Further studies are notably projected for more detailed exploration of the relationships between urban form and lifestyles. They are due to be published in the near future.

Reference

Emelianoff C, Mor E, Dobre M et al (2012) Modes de vie et empreinte carbone. Prospective des modes de vie en France et empreinte carbone. Cah du Club d'ingénierie Prospect Energ Environ 21:39–125

Appendix

This appendix gives a more in-depth description of the methodology used for estimating the energy-related performances of the current situation and the scenarios.

A.1 Buildings

This category addresses energy linked with residential and office buildings. It has been divided into two sub-categories. The first lists the energy consumed for construction of a building, whereas the second assesses the quantity of energy associated with the daily running of the building, i.e. the needs for heating, running water, ventilation, as well as for electric lighting and appliances.

For this category, requirements imposed by Minergie regulations have been used as reference values. The Minergie-A standard sets high demands in terms of comfort, especially for air and surface temperature, air humidity, indoor air pollutants, protection against external noise, preventing draughts, departure temperature for the heating circuit, and maximum ease of use. The use of renewable energy is also encouraged, as well as the minimisation of the building's grey energy and the needs of domestic electrical appliances. The Minergie-A value is set at 0 kWh/m^2. However, this can rise to 15 kWh/m^2 of total energy consumption in the form of biomass if at least 50 % of heat requirement for heating and running water is provided by solar power. The limit value for grey energy is 50 kWh/m^2. Lighting performance must be optimal and household appliances should be the most efficient class available.

The **Minergie-P** label demands in terms of comfort are high especially regarding air temperature and surfaces temperature, air humidity, interior air pollutants, protection against exterior noise, prevention of draughts but also initial temperature for departure of the heating circuit, ease of operation, information for the user, cost efficiency as well as aesthetics. At least 20 % of energy needs for heating water must be covered by renewable energy. Basic requirement for the building shell is 15 kWh/m^2, and the limit value for total energy consumption is set at 30 kWh/m^2. Specific thermal power for air heating must not exceed 10 W/m^2.

S. Lufkin et al., *Strategies for Symbiotic Urban Neighbourhoods*,
SpringerBriefs in Applied Sciences and Technology,
DOI 10.1007/978-3-319-25610-8

In Minergie-P constructions, the best conditions allowing low electricity consumption must be created. On the one hand, this demands optimal lighting and lamps and, on the other, exclusive use of A-grade efficiency household appliances.

Minergie buildings must meet the following requirements: the building shell should be well-insulated thermally and air-tight (basic requirement for new constructions), aeration system for air renewal must be controlled, low energy consumption, respect for Minergie limit value, limitation of additional costs, etc.

For all categories of buildings, their high level of airtightness presupposes a controllable exterior air duct, essential for comfort, with or without heat recovery. Non-controlled aeration (manual) through windows is insufficient to meet Minergie standards. For domestic water-heating requirements higher than 10 kWh/m^2 of energy reference area (AE), 20 % of running-water needs should be covered by renewable energy. Moreover, this is valid for all building categories: waste heat must absolutely be recycled. The ratio between the area of window AW and the energy reference area AE (AW/AE) must be smaller than or equal to 30 %.

A.2 Current Situation

In the flow analysis reflecting the current situation, primary building-linked energy consumption is assessed at 1800 W/pers (Novatlantis 2011). Surface per resident is 60 m^2 per person per flat and 37 m^2 per job per office. The figure for offices remains stable for all scenarios, whereas apartment surfaces vary.

A.3 Technological Scenario

The technological scenario meets the Minergie-A standard, i.e. 50 kWh/m^2 for construction, 15 kWh/m^2 for operating the heating, running water and ventilation sub-categories, and 45 kWh/m^2 for lighting and household appliances. For building operation, total consumption comes to 226 W/pers of total energy.

A.4 Behavioural Scenario

In the behavioural scenario only the Minergie standard was enforced. Grey energy generated by a building made from local wood is calculated on the basis of the theoretical mass balance carried out for a typical, multi-level residential building with wooden structure and surfaces (see Table A.1). These figures are from a study of wooden building constructions made in New Zealand (John et al. 2009). They remain relevant since they only refer to construction mode; eco-balance figures according to materials developed for the Swiss context (KBOB 2012) are applied subsequently.

Table A.1 Mass balance of a wooden building (John et al. 2009)

Materials	Mass [t]
Lean concrete	53.00
Heavy-load concrete	1316.00
Steel	29.07
Glass	31.94
Hardwood	130.82
Coniferous	22.80
Laminated wooden beams	343.94
Plywood	64.79
Wood fibreboard	70.62
Aluminium	1.06
Paint	1.11
Glass wool	8.42

This provides a grey energy amount of 4.12 W/m^2 (36 kWh/m^2). For the heating, running water and ventilation sub-categories the total consumption for reference is 38 kWh/m^2. This figure rises to 45 kWh/m^2 for lighting and household appliances. For building operation, the total consumption of final energy comes to 324 W/pers.

A.5 Symbiotic Scenario

The symbiotic scenario applies the intermediary Minergie-P standard. 90 MJ/m^2 (25 kWh/m^2) are allocated to the area of construction. For heating, running water and ventilation sub-categories, final consumption is 30 kWh/m^2. As in the other scenarios, this figure rises to 45 kWh/m^2 for lighting and household appliances. For building operation, the total final energy consumption comes to 307 W/pers.

A.6 Hybrid Scenario

Thanks to the hybrid solutions it offers (Minergie A for building operation, Minergie P for construction), the integrated scenario obtains the most efficient result. For building operation, the total final energy consumption comes to 160 W/pers.

The graphs in Figs. A.1, A.2 and A.3 highlight total primary energy, non-renewable primary energy and global warming potential for the four scenarios concerning the different sub-categories of the field *Buildings* (construction and operation). For the technological scenario, the figure of 226 W/pers of final energy consumed rises to 642 W/pers of primary energy due to the supply source. Wind-powered energy, in spite of its primary energy factor (PEF) of 1.33 MJ/MJ

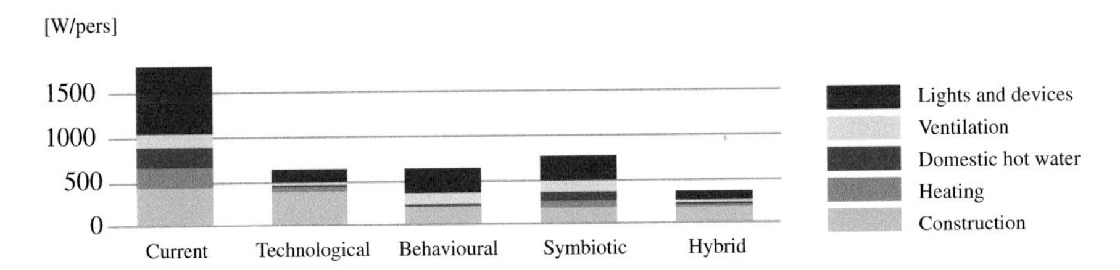

Figure A.1 Total primary energy consumption for the different sub-categories in the field of *Buildings*

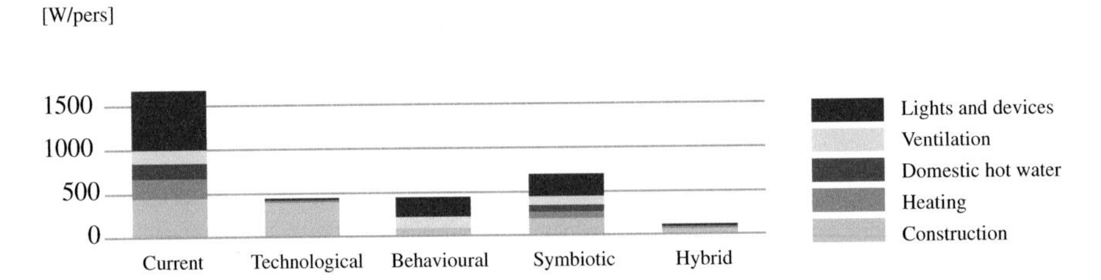

Figure A.2 Non-renewable primary energy consumption for the different sub-categories in the field of *Buildings*

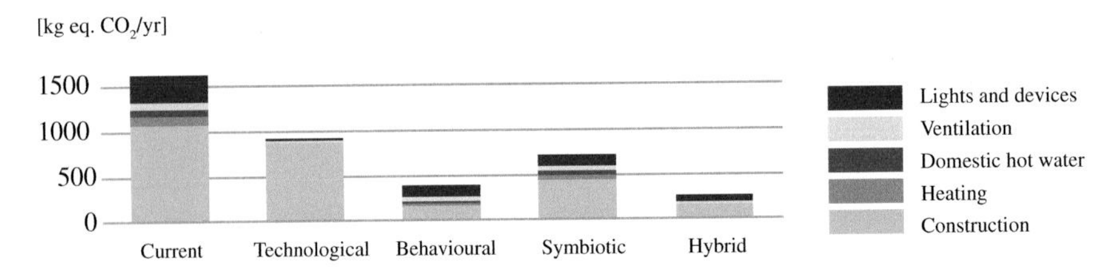

Figure A.3 Global warming potential for the different sub-categories in the field of *Buildings*

for electricity production, significantly increases primary energy consumption. It can be seen here that the proportion used for construction represents almost two thirds of total consumption.

Thanks to lightweight construction using local wood, the reverse is true for the behavioural scenario. Nevertheless, the figure rises from 324 W/pers of final energy to 651 W/pers of primary energy. This is mainly due to the fact that the demand in electricity is met by the Mix consumer CH (2.97 [MJ/MJ]). The 'greediest' scenario is the symbiotic one: here the figure rises from 307 W/pers of final energy to 778 W/pers of primary energy. This is also due to the supply source used (also the Mix CH for electricity). For building operation, the integrated scenario requires only 162 W/pers of primary energy thanks to renewable supply sources (geothermal power, solar power and cogeneration of biogas, if necessary). The total consumption of primary energy in this scenario is slightly lower than 360 W/pers.

A.7 Mobility

A.7.1 Technological scenario

In the technological scenario, 50 % of users resort to hydrogen-powered cars for travelling by individual motorised transport (IMT). The remaining users travel in conventional vehicles. Figures for hydrogen fuel cell cars were calculated on the basis of the following hypotheses: 81.25 % of the electricity supply comes from geothermal power, 12.33 % from photovoltaic (PV) panels on the roof and 6.42 % from PV panels on the façade. Table A.2 provides a synopsis of values and distances associated with the different means of transport:

A.7.2 Behavioural Scenario

Figures associated with the different types of transport for the behavioural scenario are summed up in Table A.3:

Table A.2 Distances and primary energy factors of the different types of transport for the technological scenario

Transport	km/p.year	Total PEF	Non-ren. PEF	GWP
Car (50 % H_2; 50 % normal)	9957	1610	583	1076
Plane	5238	317	313	639
Train (national)	2976	110	49	45
Train (international)	254	4.37	1.9	2.08
Bus, tramway	597	31.61	31.23	62.1
Other	712	75.2	73.4	138.13
Tourism	1420	150	146.4	275.6
Soft mobility	754	0	0	0

Table A.3 Distances and primary energy factors of the different types of transport for the behavioural scenario

Transport	km/p.year	Total PEF	Non-ren. PEF	GWP
Car (50 % H_2; 50 % normal)	8022	847	826	1556
Plane	3579	186	185	386.5
Train (national)	4062	150	66.5	61.34
Train (international)	1913	32.94	14.32	15.65
Bus, tramway	777	41.2	40.7	80.9
Other	712	75.2	73.4	138.13
Tourism	1420	150	146.4	275.6
Soft mobility	847	0	0	0

Table A.4 Distances and primary energy factors of the different types of transport for the symbiotic scenario

Transport	km/p.year	Total PEF	Non-ren. PEF	GWP
Car (50 % H_2; 50 % normal)	9957	658	643	1007
Plane	3889	235.55	233.1	474.5
Train (national)	2976	110.41	48.69	44.94
Train (international)	1602	27.6	12	13.11
Bus, tramway	597	31.61	31.23	62.1
Other	712	75.2	73.4	138.13
Tourism	1420	150	146.4	275.6
Soft mobility	754	0	0	0

A.7.3 Symbiotic Scenario

Table A.4 summarizes the values associated with the different types of transport in the symbiotic scenario. It was decided that 50 % of users use biodiesel fuel which has a positive impact of Individual motorized mobility-linked consumption.

A.7.4 Hybrid Scenario

The figures associated with the different types of transport in the hybrid scenario are summed up in Table A.5.

The graphs in Figs. A.4, A.5 and A.6 highlight total primary energy, non-renewable primary energy and global warming potential for the four scenarios concerning the different sub-categories of the field of *Mobility*.

Table A.5 Distances and primary energy factors of the different types of transport for the hybrid scenario

Transport	km/p.year	Total PEF	Non-ren. PEF	GWP
Car (50 % H_2; 50 % normal)	8022	710	305	678
Plane	3889	235	233	474
Train (national)	4062	150	66.5	61.3
Train (international)	1602	27.6	12	13.11
Bus, tramway	777	41.1	40.6	80.8
Other	712	75.2	73.4	138.13
Tourism	1420	150	146.4	275.6
Soft mobility	847	0	0	0

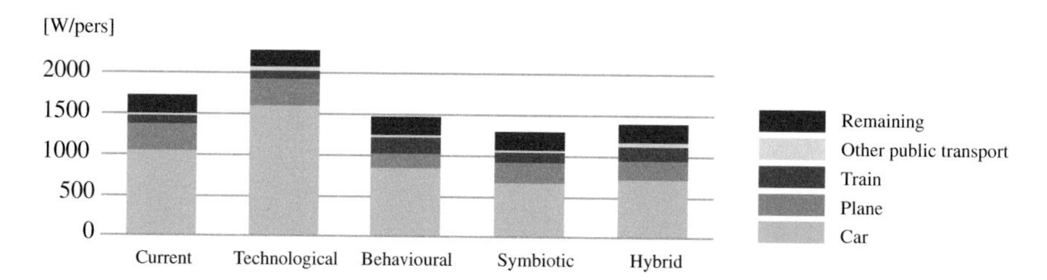

Figure A.4 Total primary energy consumption for the different sub-categories in the field of *Mobility*

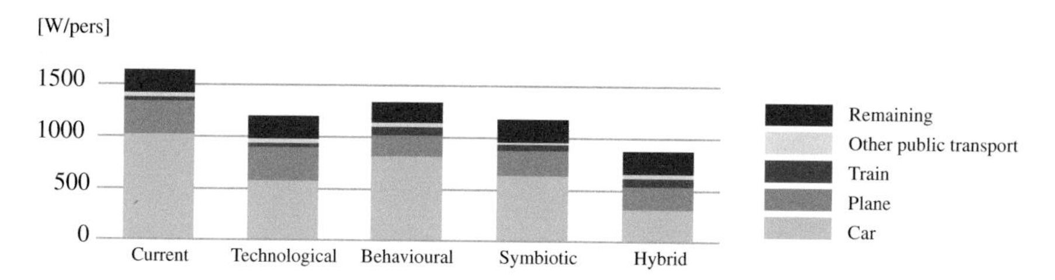

Figure A.5 Non-renewable primary energy consumption for the different sub-categories in the field of *Mobility*

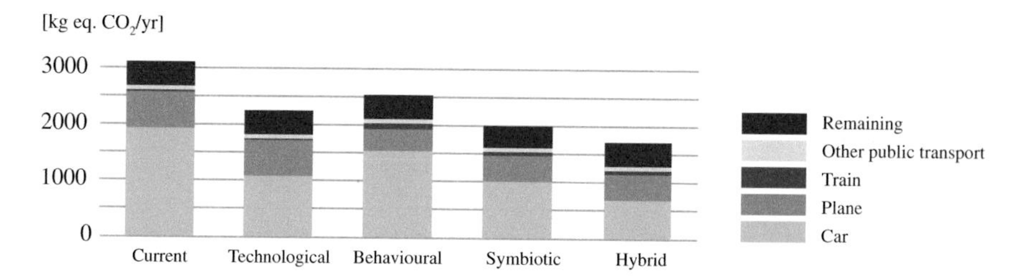

Figure A.6 Global warming potential for the different sub-categories in the field of *Mobility*

A.8 Infrastructures

Facilities listed in the Gare-Lac sector are:

- The *Centre St-Roch*, which incorporates the *HEIG-vd*, the *Bureau des Affaires Sociales*, the *Office Régional de Placement*, a number of cantonal, communal and private schools, several medical and paramedical surgeries including the *CINOV Centre d'Imageries Médicales*, various private and public offices and other craft and industrial activities as well as a restaurant, a fitness club, a dancing school and a bowling alley.
- A sports centre housing a municipal swimming pool and a skating rink.

Table A.6 Final consumption and agents used for the energy supply for facilities according to the different scenarios

Final consumption [W]		Current	Technological	Behavioural	Symbiotic	Hybrid
St Roch centre						
Thermal	102,106	Gas + fuel	Geothermal	Biogas cogeneration	Treated water + CHP	Heat pump wastewater
Electrical	364,663	Mix consumer CH	Wind	Mix consumer CH	Mix consumer CH	Wind
Swimming pool and skating ring						
Thermal	176,941	Gas	Geothermal	Biogas cogeneration	Treated water + CHP	Heat pump wastewater
Electrical	37,126	Mix consumer CH	Wind	Biogas cogeneration	Mix consumer CH	Wind
Tennis						
Thermal	10,502	Gas	Geothermal	Biogas cogeneration	Treated water – CHP	Heat pump wastewater
Electrical	2747	Mix consumer CH	Wind	Biogas cogeneration	Mix consumer CH	Wind
WPP						
Thermal	228,311	Biogas cogeneration	Self-production	Self-production	Self-production	Self-production
Electrical	136,986	Biogas cogeneration + Mix consumer CH	Biogas cogeneration + wind	Biogas cogeneration	Biogas cogeneration + mix consumer CH	Biogas cogeneration + wind
CFF workshop						
Thermal	790,636	Gas + fuel + wood	Geothermal	Biogas cogeneration	Treated water + CHP	Heat pump wastewater
Electrical	319,254	Mix consumer CH	Wind	Mix consumer CH	Mix consumer CH	Wind

- A tennis club
- The Water Purification Plant (WPP)
- The CFF railway workshops

The total current consumption of these various infrastructures, in terms of thermal energy and electricity, were provided. Different agents were then used for energy supply, depending on the characteristics of each scenario. The primary energy necessary for these different infrastructures was then assessed on this basis (see Table A.6).

The graphs in Figs. A.7, A.8 and A.9 highlight total primary energy, non-renewable primary energy and global warming potential for the four scenarios concerning the different sub-categories of the field of *Infrastructures.*

A.9 Food

With regard to the field of food, which includes agriculture, processing, packaging and distribution, current trends show an average consumption of some 750 W/pers. On the basis of recent research work (Jungbluth and Itten 2012), a certain number of categories presenting a potential for primary energy consumption reduction according to eating behaviour have been identified. They include (see table): vegetarian, local, organic, seasonal, dietetic (without excess), responsible (without waste), and unrefined (avoiding refined products such as coffee, chocolate, alcohol) nutrition. In order to calculate consumption by neighbourhood residents, different objectives were then allocated to certain groups of the population according to the ideological 'toolkit' chosen for the scenario analysed. Table A.7 illustrates this approach:

The graphs in Figs. A.10, A.11 and A.12 highlight total primary energy, non-renewable primary energy and global warming potential for the four scenarios concerning the field of *Food* (local and imported).

A.10 Goods and services

In Switzerland today, power linked with the consumption of goods and services is rated at around 750 W/pers (Novatlantis 2011). *Goods* notably include energy expenditure linked with cars, clothes, shoes, electrical household appliances, furniture, telephony, insurances, information technology, televisions, computers and flat screens, etc. *Services* include: administration, cafés, hotels and restaurants, shops, businesses, elementary, secondary and high schools, health, sport, leisure, culture, etc.

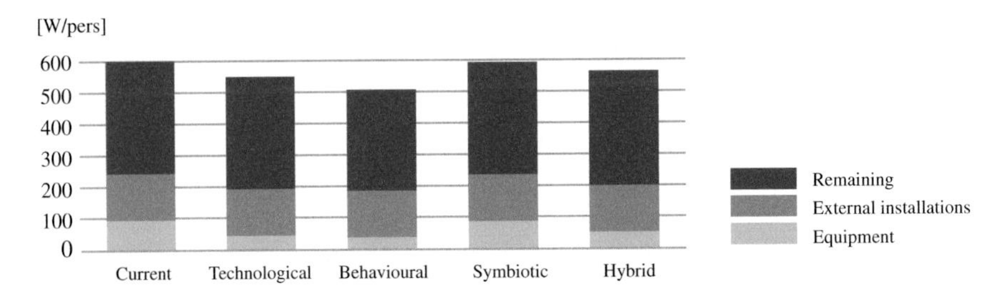

Figure A.7 Total primary energy consumption for the different sub-categories in the field of *Infrastructures*

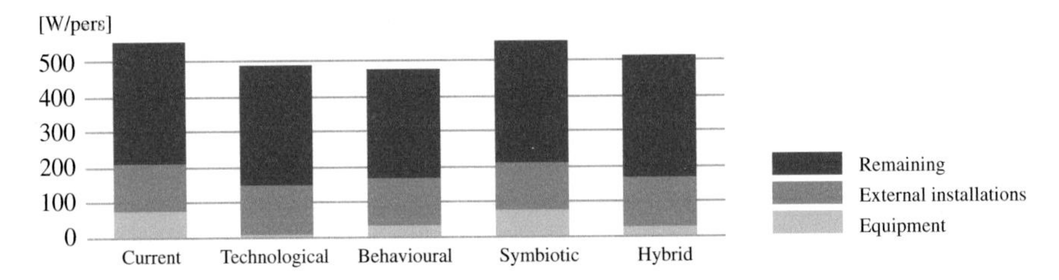

Figure A.8 Non-renewable primary energy consumption for the different sub-categories in the field of *Infrastructures*

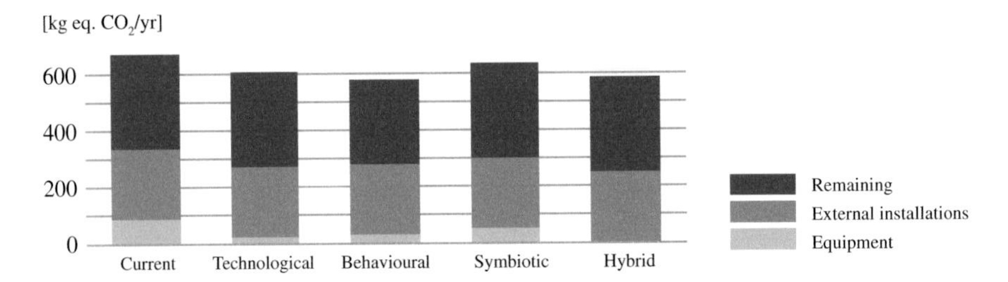

Figure A.9 Global warming potential for the different sub-categories in the field of *Infrastructures*

Table A.7 Potential for reducing energy consumption for the sub-categories in the field of food and percentages of the population involved in the effort

Food	Reduction (%)	Percentage of population involved (%)			
		Technological	Behavioural	Symbiotic	Hybrid
Vegetarian	−35.0	10	70	10	30
Local	−1.0	5	70	70	50
Organic	−6.2	5	70	0	50
Seasonal	−2.0	5	70	70	50
Dietetic	−4.9	0	0	70	50
Responsible	−10.1	0	70	50	30
Unrefined	−12.0	0	70	20	10
Total	−71.2	0	−46	−13	−22

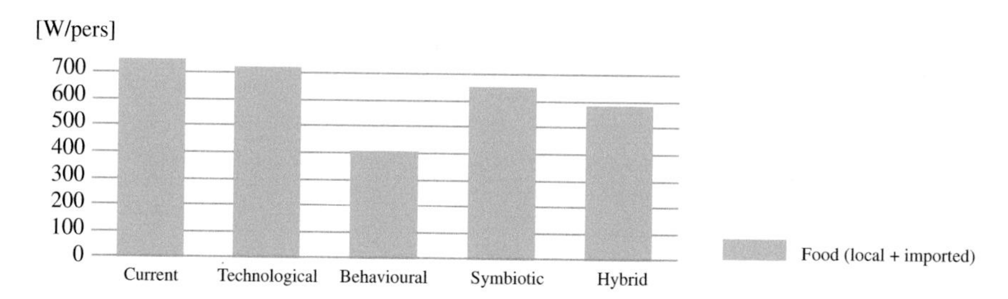

Figure A.10 Total primary energy consumption for the field of *Food*

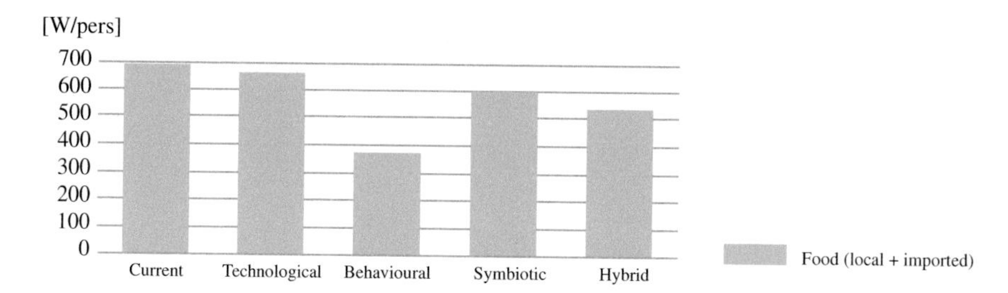

Figure A.11 Non-renewable primary energy consumption for the field of *Food*

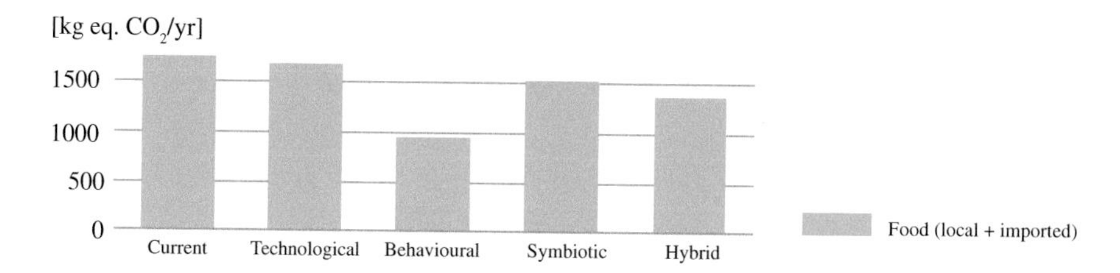

Figure A.12 Global warming potential for the field of *Food*

As was the case with research into eating habits, the lack of reliable data available in this field has meant that sub-categories were not really relevant. This category's total primary energy, non-renewable primary energy and Global Warming Potential are therefore assessed in proportion to the average reference value depending on the hypotheses of each scenario.

Table A.8 illustrates this approach:

Table A.8 Consumption linked with *Goods and services* in the different scenarios, depending on the three indicators

	Current	Technological	Behavioural	Symbiotic	Hybrid
Total primary energy [W/pers]	750	750	638	675	750
Non-renewable primary energy [W/pers]	750	750	638	675	750
Global warming potential [kg eq CO_2/year]	1012	1012	860	911	1012
Reduction (%)	–	0	−15	−10	0

The graphs in Figs. A.13, A.14 and A.15 highlight total primary energy, non-renewable primary energy and global warming potential for the four scenarios concerning the field of *Goods and services.*

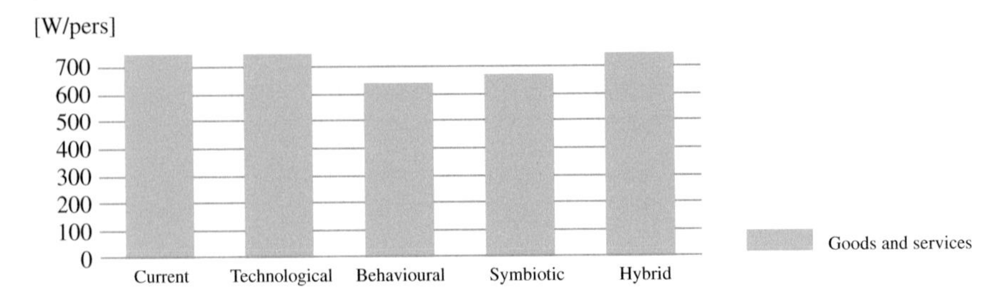

Figure A.13 Total primary energy consumption for the field of *Goods and services*

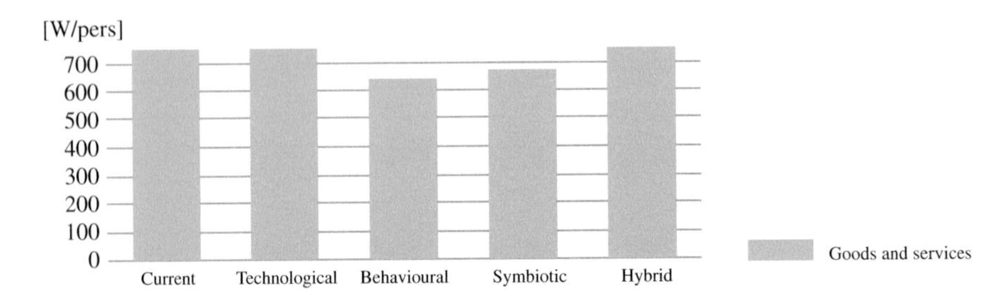

Figure A.14 Non-renewable primary energy consumption for the field of *Goods and services*

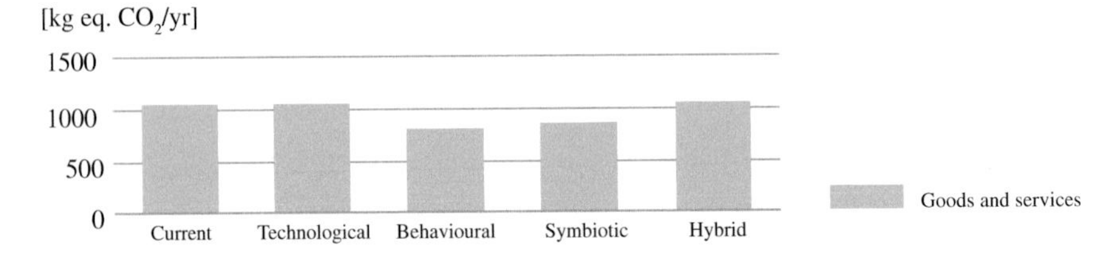

Figure A.15 Global warming potential for the field of *Goods and services*

References

John S, Nebel B, Perez N, Buchanan A (2009) Environmental impacts of multi-storey buildings using different construction materials. Christchurch.

Jungbluth N, Itten R (2012) Umweltbelastungen des Konsums in der Schweiz und in der Stadt Zürich: Grundlagendaten und Reduktionspotenziale. Zurich.

KBOB (2012) Données des écobilans dans la construction 2009/1. Bern.

Novatlantis (2011) Vivre plus légèrement. Vers un avenir énergétique durable: l'exemple de la société à 2000 Watts. Villigen.